Contents

Command words

The command words that are used in questions give you a good idea as to what the examiner is looking for. A number of them are shown below with their meaning. These are grouped in level of difficulty.

Identify   From the table, map or stimulus material choose the correct feature or factor.

State   Mention briefly; a word may suffice but it is better if you use a whole sentence.

Define   Give the meaning of the term.

Describe   Give the main characteristics; this may be a pattern or the features of a particular landform. No explanation is required but certain descriptors (such as dispersed, clustered or a description of scale) may be appropriate.

Outline   Give the main characteristics or reasons for a particular geographical phenomenon.

Explain   Give reasons how a process or a pattern functions. There will be no marks for description unless the question says 'describe and explain'.

Account for   Another way of asking for reasons.

Suggest why   As above; this implies that there may be more than one correct answer depending on the case studies or examples used as support.

Using examples or case studies   There is a clear request for examples and/or case studies. Normally about half of the marks are reserved for the named and located details.

With reference to a place you have studied   Again, a clear request for real information about real places.

Compare   In what ways are two or more phenomena similar? What have they got in common?

Contrast   In what ways are two or more phenomena different? How do they differ?

Examine   Generally requires some description and explanation about the nature of a process or feature.

Assess   Compare and contrast the importance of some factor, process, or feature relative to others.

To what extent   A question that requires an answer! It requires an evaluation (assessment) of the importance of particular factors or processes and a conclusion suggesting how important they are.

Evaluate   Contrast and state which is most important and why.

NB: L = lines; M = marks

Model answers

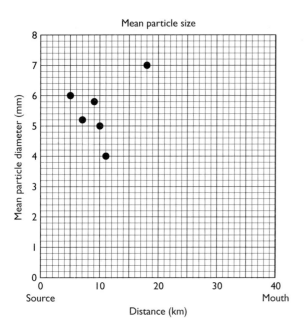

Mean particle size

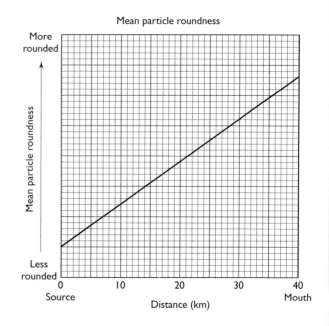

Mean particle roundness

The following questions were asked at A/S level.

Deposition by rivers

A Complete Figure A by plotting the data in Table 1. (*2M*)

B Draw a best-fit straight line on Figure A. (*1M*)

C Comment on the information shown by the best fit lines in Figures A and B. (*5L, 3M*)

D Explain the changes which occur with increasing distance from the river source in

(i) mean particle roundness (*4L, 3M*)

(ii) mean particle size. (*4L, 3M*)

E Explain why river deposits frequently show stratified bedding (*4L, 2M*)

F For a river you have studied

(i) explain how human activity may increase river deposition,

(ii) outline the various ways in which people make use of river deposits. (*12L, 6M*).

Table 1

Distance (km)	Mean particle diameter (mm)
15	3.0
20	3.5
21	3.1
22	2.6
27	3.0
32	2.5

Adapted mark scheme

A Lose ½ mark per error, and round down, e.g. 0 errors = 2 marks, 1 error = 1 mark, 2 errors = 1 mark, 3 errors = 0 marks, 4 errors = 0 marks. (*2*)

B The line must be straight *not* curved. It should pass approximately through the middle. (*1*)

C With increasing distance downstream particle size decreases (*1M*) and particles become rounder (*1*). Additional mark for comment on correlation i.e. inverse or negative relationship between the two (*1*) (*3×1*)

D (i) more attrition so pebbles become rounder (*1*); edges knocked off by longer periods of particles knocking against each other (*1*); more of the smaller, rounder pebbles are carried downstream (*1*); marks awarded for breadth and/or depth of explanation. (*3×1*)

(ii) longer period of attrition so particles become smaller (*1M*); carrying capacity reduced so only finer particles can be carried (*1M*); award candidates that use the data, and refer to the fact that the changes in distance and particle size are not steady increases. Some comment on this, with a suggestion why, should be rewarded (*1–2M*) (*3×1*)

E Seasonal changes in velocity – high in winter, leading to a coarser load (*1M*) low in summer leading to a finer load (*1M*), or the

changes may be with distance downstream – upper course coarse load (*1*), lower course fine load.(*1*) (*2×1*)

F Maximum of *3/6* if no example is used

(i) ▨ dam construction reduces river velocity and increases deposition

▨ water extraction for industry, agriculture, tourism, etc. thereby reducing the flow of water

▨ fly-tipping (dumping) in a river to increase its load, and block the river

(ii) ▨ gravel for building industry

▨ silt for farming

▨ sand or clay for building, especially in LEDCs

Some may think of river deposits such as terraces being used for settlement, transport, etc. Reward creditable answers, but remember that examples are required. (*6*)

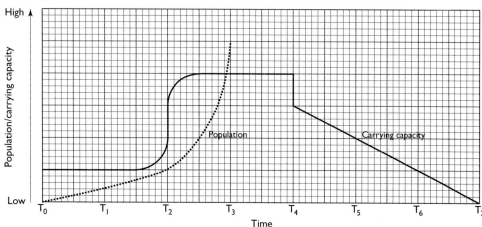

Population and resources

Study Figure C, which shows changes in population and carrying capacity over time.

A Define the term 'carrying capacity'. (*4L, 2M*)

B On a copy of Figure C, complete the graph to show how the population may change over time. (*3M*)

C Explain the relationship you have shown between population and carrying capacity for the period T3–T7. (*6L, 4M*)

D Suggest how
(i) economic factors could have increased the carrying capacity at T2
(ii) physical factors could have decreased carrying capacity at T4. (*12L, 8M*)

E For a region or country you have studied, explain how government policies have influenced population totals. (*8L, 8M*)

Mark scheme

A The number of people who can be supported by the resources of an area. (1) For two marks, there should be some reference to the idea of quality of life or standard of living. As standards of living fall, the carrying capacity has been exceeded. (2)

B and **C** are best marked together. C must justify B. B can vary depending on the line the candidate takes. For example

- population crashes following famine and disease
- population is stabilised through aid effort
- out migration may reduce population numbers
- 'overshoot' may be prevented through population policies.

NB: The question says *explain*, so candidates must say *why* the population curve changes, rather than describe the changes. (3)

D (i) Answers need to link the changes to increased carrying capacity e.g.

- new irrigation methods (1)
- use of high-yielding varieties of seed (1)
- increased use of fertiliser (1)
- hydroponics (1)
- greenhouses (1)
- better storage (1)

NB: Pupils must show how the above initiatives will lead to greater food supply, and therefore an increase in the carrying capacity. (4×1)

(ii) Any disadvantage resulting from increased production or overproduction, e.g.

- salinisation
- soil degradation
- falling groundwater levels
- evaporation losses in dams
- wind and water erosion following vegetation removal
- desertification
- resource exploitation. (4×1) (4)

(iii) Without examples the maximum mark is 4/8.

Reward candidates who use countries other than China, and reward candidates who look at pro-natalist countries, such as France, as well as those that have changed from anti-natalist to pro-natalist, such as Singapore. Countries may also operate an open-door policy and welcome immigrants into their country.

Governments may encourage couples to have more children via

- favourable maternity and paternity breaks
- free schooling and education for children
- creches in the workplace
- generous child benefit allowances
- improvements in female education
- increasing the number of women in the workforce
- running love cruises (as in Singapore) to encourage single adults to meet).

This may encourage having more than 2 children (i.e. above the replacement level) and so the total population will grow. So too will government legislation which allows immigrants to enter easily into a country.

Governments may be anti-natalist in many ways:

- a taxation system that punishes people with more than one child
- forced abortions
- forced sterilisations and vasectomies
- through the education system
- via advertising on the media
- by making life difficult for people who choose to have above a certain number of children; this can be in the workplace, the village, etc.

These methods should cause the birth rate to fall below the replacement level and so the population total should fall. (8)

1. Urbanisation and hydrology

Figure 1 shows some of the effects of urbanisation on hydrology.

Figure 1 Some effects of progressive urbanisation

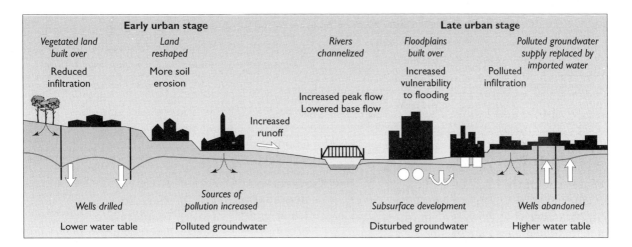

A Define the following terms:
- infiltration
- groundwater
- flood plain
- runoff. (*8L, 4M*)

B Describe and explain the impact of wells on the water table, during the early urban stage. (*4L, 3M*)

C Comment on the changes to the water table during the late urban stage. (*4L, 3M*)

D Suggest the problems that this may cause for a large urban area such as London. (*6L, 4M*)

E Describe the changes to
(i) surface water and
(ii) ground water
as urbanisation proceeds. (*5L, 4M*)

F Suggest why the quality of groundwater changes over time. (*4L, 4M*)

G In what ways would the typical storm hydrograph change between the early urban stage and the late urban stage. Use a labelled diagram in your answer. (*4L, 3M*)

2. Impact of dam construction

A The data in Figure 2 shows silt concentrations on the Nile at Gaafra before and after the construction of the Aswan High Dam.

Figure 2 Silt concentrations in the Nile before and after dam construction

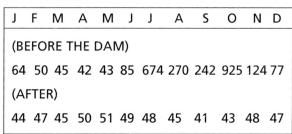

J	F	M	A	M	J	J	A	S	O	N	D
(BEFORE THE DAM)											
64	50	45	42	43	85	674	270	242	925	124	77
(AFTER)											
44	47	45	50	51	49	48	45	41	43	48	47

(i) Plot the data for silt concentration before and after the building of the High Dam. (graph paper needed) (*4M*)

(ii) Describe the changes in silt concentrations before and after the building of the High Dam. (*6L, 4M*)

Figure 3 Possible effects of dam construction on the environment

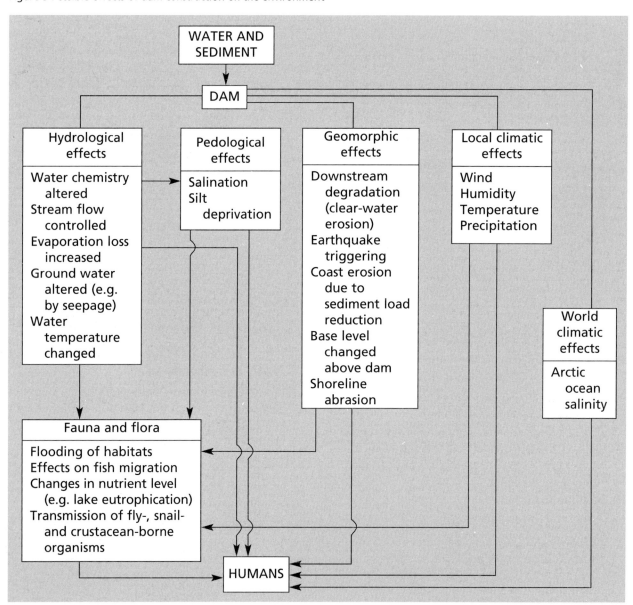

Figure 3 shows some of the possible impacts of dam construction on the environment.

B Describe and explain any *two* effects on the hydrological cycle. (*10L, 6M*)

C Suggest what is meant by the term 'clear water erosion'. (*4L, 2M*)

D Explain why clear water erosion increases below the dam. (*5L, 3M*)

E Give reasons to explain why the construction of large dams may lead to

(i) shoreline abrasion (coastal erosion) and

(ii) earthquake triggering. (*10L, 6M*)

3. Aquifers and nitrate pollution

Figures 4A, B and C show the agricultural pollution risk, the changes in pollution of groundwater over time, and major aquifers at risk from nitrate pollution.

A Describe the distribution of areas most at risk of nitrate pollution. (*5L, 4M*)

B Suggest reasons why this area is most 'at risk'. (*5L, 4M*)

C Outline the differences between confined and unconfined aquifers.

(See question 29 on pages 30–31) (*3L, 2M*)

Figure 4B shows changes in the amount of nitrate-prone subsurface water over time.

D Describe the distribution of nitrates in groundwater with depth in

(i) 1977

(ii) 1979

(iii) 1984 (*9L, 9M*)

E Comment on the implication of the trend you have shown for changes in water quality over time. (*4L, 3M*)

F Suggest ways in which nitrate levels in groundwater may be reduced. (*4L, 3M*)

Figure 4A

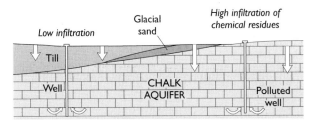

Figure 4B

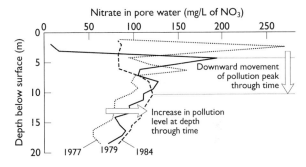

Figure 4C

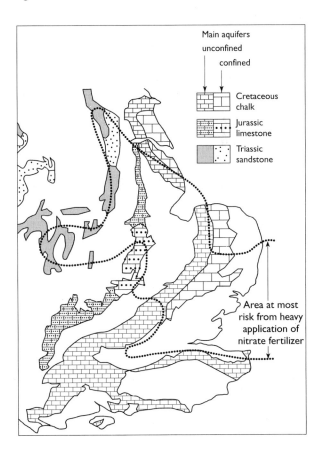

4. Water management

Figure 5 shows some impacts of water management.

Figure 5

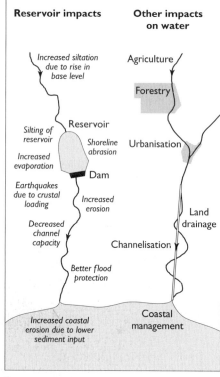

A Other than better flood protection, state three benefits of reservoir development. (*3L, 3M*)

B Explain any *two* of the effects of the reservoir above the dam. (*9L, 6M*)

C Suggest why the development of large dams may lead to increased earthquake activity. (*4L, 3M*)

D Explain briefly the term 'crustal loading'. (*4L, 3M*)

E Explain why construction of the reservoir might result in increased shoreline abrasion in the reservoir. (*4L, 3M*)

F Suggest a definition for the term 'channelisation'. (*3L, 1M*)

G Describe the changes to the river channel as a result of channelisation, as shown in Figure 5. (*4L, 3M*)

H Explain how the changes in the river channel will lead to changes in processes within the river. (*4L, 3M*)

5. Global sediment yields

Figure 6 shows average annual sediment yields.

A (i) Define the term 'discharge'. (*2L, 2M*)

   (ii) Define the term 'sediment yield'. (*2L, 2M*)

Figure 6

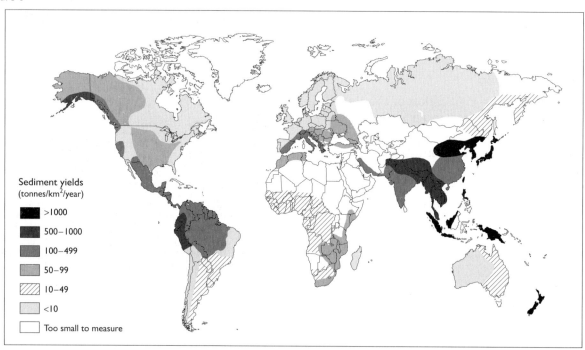

Sediment yields
(tonnes/km²/year)

- >1000
- 500–1000
- 100–499
- 50–99
- 10–49
- <10
- Too small to measure

B Study the map of global variations in sediment yield.

Describe the distribution of areas

(i) with sediment yield too low to measure, and

(ii) areas with a sediment yield of over 500 tonnes/km²/year. (*9L, 6M*)

C Suggest reasons why the British Isles has a relatively low sediment yield. (*5L, 3M*)

D With reference to examples you have studied, outline the physical and human factors that influence sediment yields. (*8L, 6M*)

E Sediment can be considered both a benefit and a problem to people. Discuss this statement with reference to examples you have studied. (*9L, 6M*)

6. Stream order

Figure 7 shows a stream ordering in a small drainage network.

Figure 7

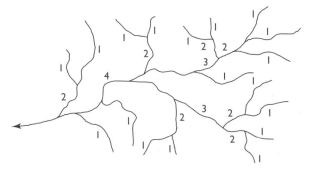

1km

A What is meant by the term 'stream order'? (*1L, 1M*)

B Describe the characteristics of a first order stream. (*3L, 2M*)

C Complete the table below to show the number of streams in each order. (*5L, 2M*)

Order	Number of streams
First	
Second	
Third	
Fourth	

D Describe the relationship between the number of streams and stream order. (*2L, 2M*)

E How does stream length vary with stream order? Use evidence to support your answer. (*4L, 3M*)

F In what ways is stream order useful for geographers? (*5L, 4M*)

G What is drainage density and how is it measured? (*3L, 3M*)

H With reference to examples that you have studied, outline the factors that influence drainage density within an area. (*12L, 8M*)

7. Meanders

A What is a 'meander'? (*2L, 2M*)

Figure 8 Straight section of the River Cole

Figure 9 Meandering section of the restored River Cole

Figure 10

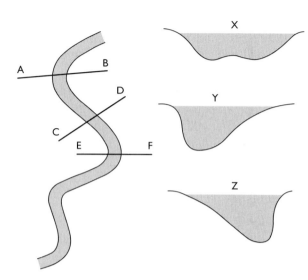

G Explain how the movement of water in a meander can lead to a variety of features. (*8L, 8M*)

8. Flood plains

Figure 11 shows a diagram of a flood plain.

Figure 11

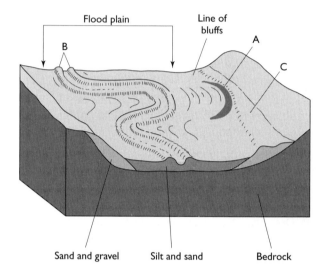

B Describe the main differences between the meandering channel in Figure 8 and the straight channel in Figure 9. (*7L, 5M*)

C Draw a diagram to show the main features found in a meander. (*6L, 3M*)

D Study Figure 10. It shows three cross-sections of a river.
Complete the table by matching the river's cross-sections X, Y and Z with the correct location cross-sections A–B, C–D and E–F. (*4L, 3M*)

Locational cross-section	Cross-section
A–B	
C–D	
E–F	

E What are 'pools and riffles'? (*4L, 2M*)

F Describe how pools and riffles are formed. (*4L, 2M*)

A Identify the features labelled A, B and C. (*3L, 3M*)

B Explain how each of the features may have been formed. (*16L, 12M*)

C Using examples suggest how landform C has influenced human activity. (*8l, 5M*)

D Outline the advantages and disadvantages of flood plains for human activity. (*8L, 5M*)

9. Hjulström curves

Figure 12 shows the Hjulström curve which displays the relationship between the velocity of water, particle size, and the work of a river.

Figure 12

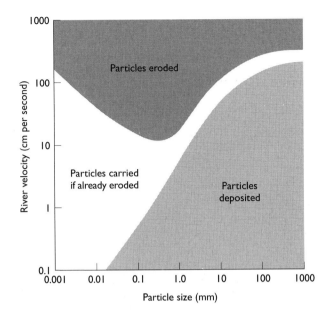

A Describe the work of a river which flows at 10 cm/second in terms of transport and deposition. (*4L, 2M*)

B Describe how water of differing velocity affects particles of 1 mm in size. (*3L, 3M*)

C Describe the relationship between river velocity and deposition. (*3L, 3M*)

D Suggest reasons for this relationship. (*3L, 2M*)

E Describe the relationship between river velocity and erosion. (*3L, 3M*)

F Suggest reasons to explain this relationship. (*5L, 3M*)

G Comment on how the work of a river may vary seasonally. (*5L, 3M*)

H With reference to a river that you have studied, describe and suggest reasons for the changes you would expect to see in the size, shape and amount of load carried from the source to the mouth. (*9L, 6M*)

10. Urban hydrographs

Figure 13 shows the impact of urbanisation on a flood hydrograph.

A Calculate the peak flow:

(i) before urbanisation

(ii) after urbanisation. (*2L, 2M*)

B Determine the time lag in days:

(i) before urbanisation:

(ii) after urbanisation: (*2L, 2M*)

C Describe the changes in the hydrograph that have resulted from urbanisation. (*5L, 3M*)

D Suggest reasons to explain the changes you noted in C. (*14L, 9M*)

E Explain how each of the following may affect the shape of a flood hydrograph

■ the shape of the drainage basin

■ the size of the drainage basin

■ the relief of the drainage basin. (*14L, 9M*)

Figure 13

RURAL

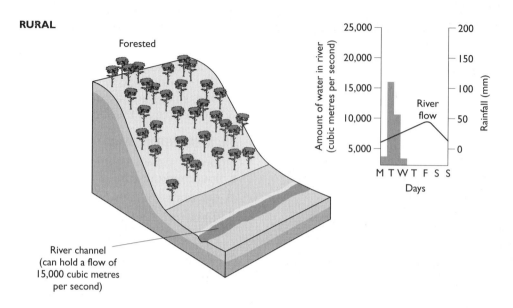

Forested

River channel
(can hold a flow of
15,000 cubic metres
per second)

URBAN

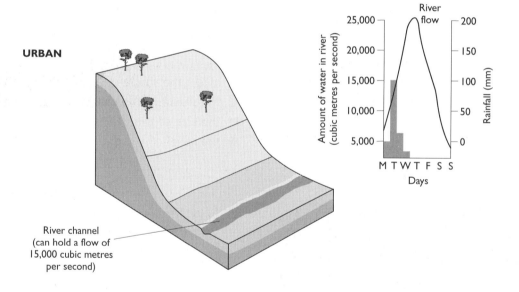

River channel
(can hold a flow of
15,000 cubic metres
per second)

Figure 14

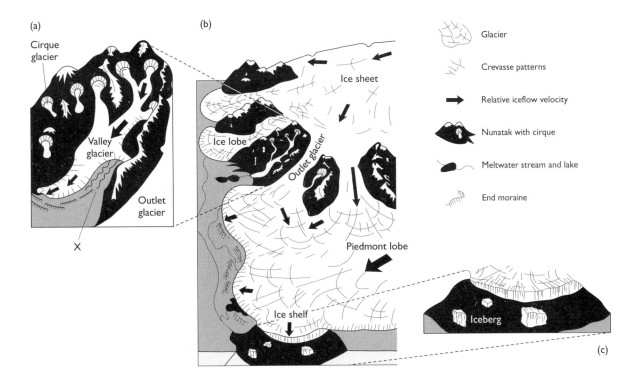

(a)

Cirque glacier

Valley glacier

Outlet glacier

X

(b)

Ice sheet

Ice lobe

Outlet glacier

Piedmont lobe

Ice shelf

(c)

Iceberg

Glacier

Crevasse patterns

Relative iceflow velocity

Nunatak with cirque

Meltwater stream and lake

End moraine

Figure 14A

11. Upland glaciation

Figure 14 shows a model of upland glaciation.

A (i) Identify four types of ice mass shown on the diagram. (*4L, 4M*)

(ii) Suggest what type of ice mass is shown in Figure 14A.

B What is a 'nunatak'? (*2L, 2M*)

C (i) Identify a likely location for a pyramidal peak. (*1*)

(ii) Describe the main characteristics of a pyramidal peak. (*5L, 3M*)

(iii) Explain how pyramidal peaks are formed. (*8L, 4M*)

D (i) Identify Feature X (*1L, 1M*)

(ii) Describe the main characteristics of Feature X. (*7L, 4M*)

(iii) Suggest how it could be formed.
(8L, 6M)

12. Cirque development

Figure 15 shows the development of a cirque over time.

A Describe the main feature of a cirque.
(6L, 4M)

B Using information in Figure 15 suggest how cirques evolve. (8L, 6M)

C In what ways do rock types influence the size and shapes of a cirque? (3L, 2M)

Altitude and orientation of the Northern Glydder cirques, Snowdonia

Cirque	Altitude (m)	Orientation (°)
Ceunant	630	040
Grainanog	610	042
Perfedd	670	051
Bual	670	059
Coch	670	060
Cywion	690	048
Clyd	810	073
Idwal	630	038
Cneifion	790	020
Bochlwyd	780	025

D (i) Work out the mean altitude of cirques in the Glydder. (2M, 2L)

(ii) Work out the mean aspect of cirques in the region. (2M, 2L)

E What other factors might be important in cirque development? (5L, 3M)

F Briefly explain how *aspect* may affect cirque development. (5L, 4M)

G Comment on the likely relationship between altitude and cirque development. (5L, 4M)

Figure 15

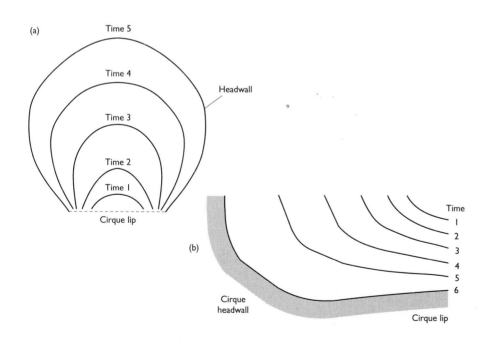

13. Glacial systems

Figure 16A shows the glacial system for an ice sheet and a valley glacier.

Figure 16A

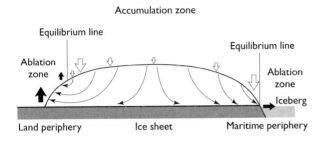

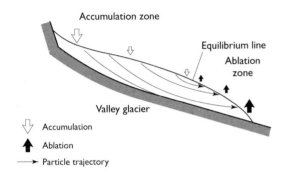

Accumulation

Ablation

Particle trajectory

A Name two inputs into the system which lead to the accumulation of ice. (*2L, 2M*)

B Identify two outputs from the system which cause ablation. (*2L, 2M*)

C Explain how snow eventually forms ice. (*4L, 3M*)

D Comment on the differences in the ablation zone for the ice sheet and the glacier. (*8L, 6M*)

Figure 16B shows annual variations in accumulation and ablation.

Figure 16B

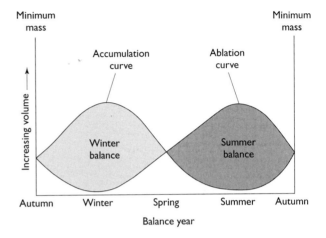

E Describe the variations in accumulation and ablation over the course of a year. (*4L, 3M*)

F Suggest reasons for the variations outlined in E. (*6L, 4M*)

G Suggest how seasonal variations in the glacial system affect human activities. (*9L, 5M*)

14. Glacial retreat

Figure 17 shows a retreating glacier.

Figure 17

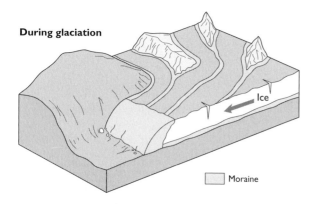

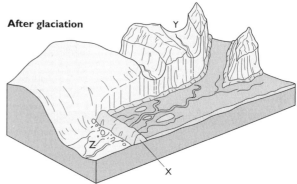

A What is the evidence to suggest that the glacier is retreating? (*3L, 2M*)

B What are the factors that may cause the glacier's snout to retreat? (*7L, 5M*)

C Identify the two types of moraine, X and Y. (*2L, 2M*)

D Describe the main characteristics of X. (*4L, 3M*)

E Suggest ways in which glacial deposits can be used to determine the direction of glacier flows. (*7L, 5M*)

Area Z on the diagram represents an outwash plain or sandur.

F Describe, and account for, the nature of the deposits on Z. (*6L, 4M*)

G Outline the problems associated with the exploitation of meltwater deposits. (*6L, 4M*)

15. Periglacial environments

Figure 18 shows the distribution of freeze-thaw over a year in different periglacial environments.

A Define the term 'periglacial'. (*2L, 2M*)

B Describe the main conditions associated with periglacial climates. (*3L, 3M*)

C Explain the process of freeze-thaw weathering. (*3L, 3M*)

D Outline the factors that increase susceptibility to freeze-thaw weathering. (*4L, 3M*)

E Which of the stations had (i) most and (ii) least cycles of freeze-thaw over the year? (*3L, 2M*)

F Describe the annual variations in temperature and freeze-thaw conditions at Sonnblick in the Alps. (*4L, 3M*)

G Suggest reasons to explain this pattern. (*6L, 4M*)

H To what extent is it possible to manage cold environments? Use examples to support your answer. (*10L, 5M*)

Figure 18

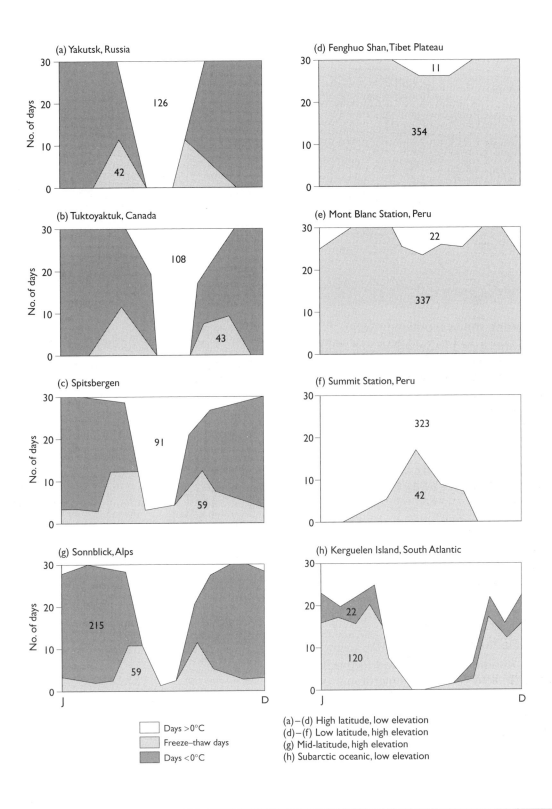

(a) Yakutsk, Russia

(b) Tuktoyaktuk, Canada

(c) Spitsbergen

(g) Sonnblick, Alps

(d) Fenghuo Shan, Tibet Plateau

(e) Mont Blanc Station, Peru

(f) Summit Station, Peru

(h) Kerguelen Island, South Atlantic

Days >0°C

Freeze–thaw days

Days <0°C

(a)–(d) High latitude, low elevation
(d)–(f) Low latitude, high elevation
(g) Mid-latitude, high elevation
(h) Subarctic oceanic, low elevation

16. Water flow in periglacial environments

Figure 19 shows a periglacial landscape in Southern Iceland.

Figure 19

The photograph shows a braided river.

A Draw an annotated (labelled) sketch diagram to show the main characteristics of a braided river. (*8L, 3M*)

B Describe the main characteristics of the river. (*3L, 3M*)

C Suggest reasons why the river has these characteristics. (*4L, 4M*)

D Suggest how the load the river carries will change with distance away from the glacier. (*4L, 4M*)

Figure 20 shows a hydrograph for stream flow in periglacial areas.

E Describe the pattern of river flow in the Arctic nival regime. (*5L, 4M*)

F Suggest why there was

(i) no flow in winter,

(ii) variable flow in July. (*5L, 4M*)

G Outline possible hazards related to the flow of water in cold periglacial environments. (*6L, 3M*)

Figure 20

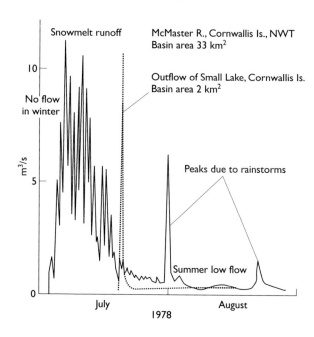

17. Fluvioglacial landscapes

Figure 21 Glacial retreat

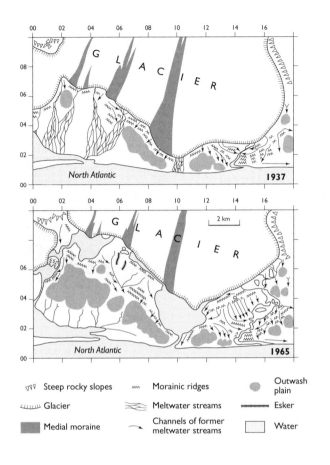

	Steep rocky slopes	⁀	Morainic ridges	●	Outwash plain
	Glacier	⧓	Meltwater streams	▬	Esker
▰	Medial moraine	↝	Channels of former meltwater streams	▭	Water

A Describe the changes in the ice mass as shown in Figure 21. (*6L, 4M*)

B Comment on the changes that have happened to the outwash plain. (*6L, 4M*)

C On a copy of Figure 21 identify the likely examples of terminal moraine, lateral or medial moraine, and recessional moraine. (*3M*)

D Comment on the location of the esker in Figure 21. (*4L, 3M*)

E What is an 'esker'? (*4L, 2M*)

F Describe how eskers are formed. (*6L, 4M*)

G Suggest the likely problems for human activities in the landscape shown in Figure 21. (*9L, 5M*)

Figure 22A

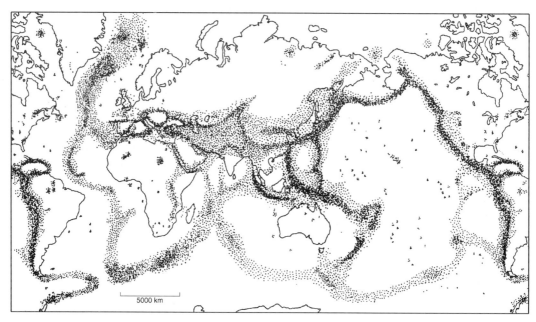

18. Hazards related to earthquakes

Figure 22 shows the global distribution of earthquakes and Figure 23 shows some of the hazards related to an earthquake.

A Briefly describe the global pattern of earthquakes. (*6L, 4M*)

B Suggest reasons for the pattern show in Figure 22A. (*7L, 4M*)

C Suggest meanings for the terms 'liquefaction', 'focus' and 'epicentre'. (*6L, 6M*)

D Describe the pattern of earthquakes in the UK 1979–90, as shown in Figure 22B. (*4L, 4M*)

E Using examples, outline the factors that cause the impact of an earthquake to vary from place to place. (*12L, 7M*)

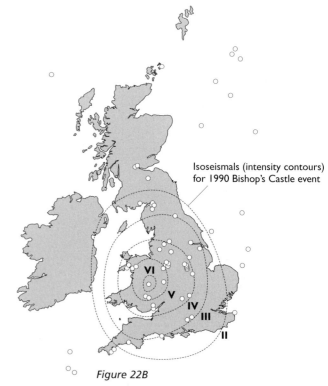

Isoseismals (intensity contours) for 1990 Bishop's Castle event

Figure 22B

Figure 23

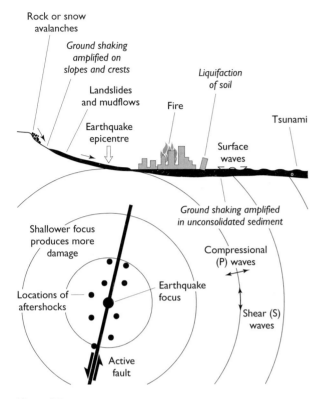

Figure 24

19. Volcanic hazards

Figure 24 shows the distribution of the world's active volcanoes and Figure 25 shows some of the impacts of a volcano, and some of the changes that happen before a volcano erupts.

A Describe the distribution of volcanoes as shown in Figure 24. (*5L, 4M*)

B Suggest contrasting reasons for the volcanoes at X, Y and Z. (*6L, 6M*)

C Describe briefly three hazards associated with volcanic eruptions. (*9L, 6M*)

D Why, despite these hazards, do people continue to live in volcanic areas? Use examples to support your answer. (*6L, 4M*)

E To what extent is it possible to manage the hazards related to volcanoes? Use examples to support your answer. (*9L, 5M*)

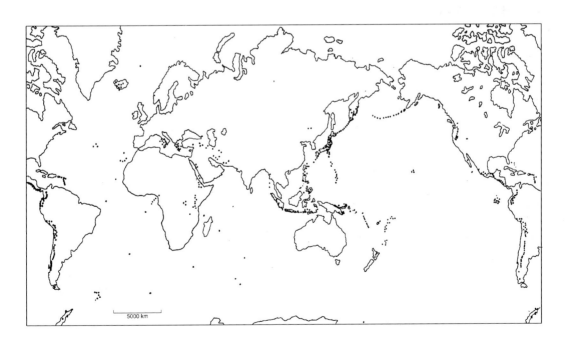

Figure 25

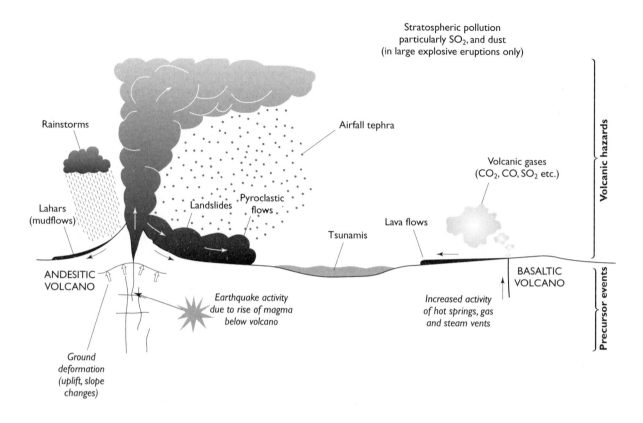

Stratospheric pollution
particularly SO_2, and dust
(in large explosive eruptions only)

Rainstorms

Airfall tephra

Volcanic gases
(CO_2, CO, SO_2 etc.)

Lahars
(mudflows)

Landslides Pyroclastic
flows

Lava flows

Tsunamis

**ANDESITIC
VOLCANO**

*Earthquake activity
due to rise of magma
below volcano*

*Increased activity
of hot springs, gas
and steam vents*

**BASALTIC
VOLCANO**

Volcanic hazards

Precursor events

*Ground
deformation
(uplift, slope
changes)*

20. The evolution of the Canary Islands

Figure 26 shows the evolution of the Canary Islands.

A Which was the first island to be formed, and when? (*2L, 2M*)

B Approximately when was Tenerife formed? (*1L, 1M*)

C Describe the general relationship between age of island and distance from Africa. (*3L, 2M*)

D Describe how plate movement has changed over time. (*5L, 3M*)

Figure 26

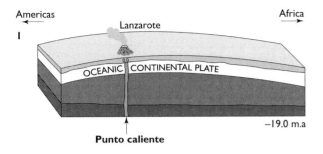

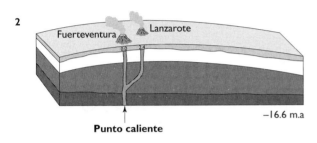

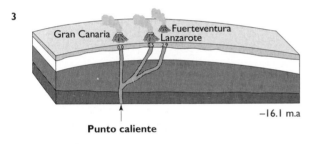

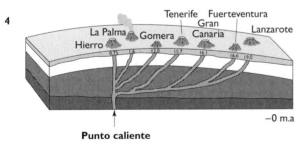

Figure 27 shows an alternative model of the evolution of the Canaries.

Figure 27

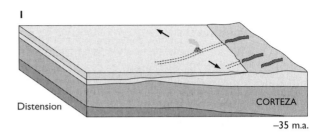

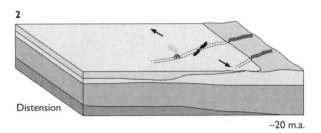

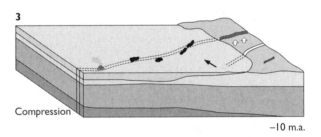

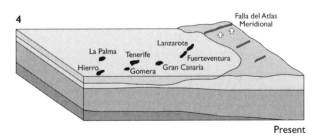

E When were
 (i) Fuerteventura and
 (ii) Gran Canaria formed? (*2L, 2M*)

F Which was the last of the Canary Islands to be formed? (*1L, 1M*)

G What is the hot spot? (*3L, 2M*)

H Explain how hot spot activity might have influenced the development of the Canary Islands. (*4L, 3M*)

Figure 28

I Suggest ways in which volcanic activity might benefit the Canary Islands. (*6L, 4M*)

J Outline the hazards associated with volcanic activity. (*8L, 5M*)

21. A model of vulcanicity

Figure 28 shows a model of vulcanicity.

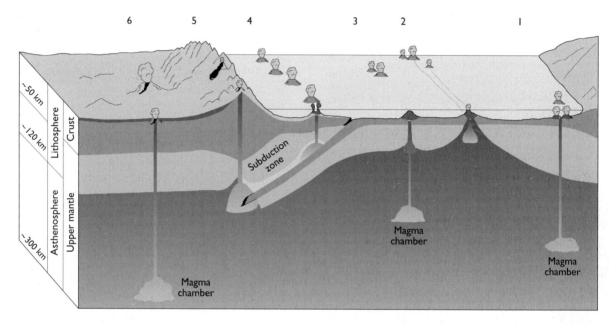

A Choose the labels for sites 1–6 (one label will be used twice)
- continental hot spot
- oceanic hot spot
- island arc
- fold mountains
- mid-ocean ridge (*3M*)

B Suggest how the magma of 2 could differ from that at 5. (*6L, 4M*)

C Identify the type of tectonic plate boundary at 2 and 5. (*2L, 2M*)

D Briefly describe the processes that are taking place at location 2. You can use a labelled diagram to help you. (*7L, 4M*)

E Compare the nature of the oceanic crust with that of the continental crust. (*6L, 4M*)

F Describe and account for the land forms associated with subduction zones. (*8L, 4M*)

G Explain why subduction zones are associated with deep focus earthquakes. (*6L, 4M*)

22. Tectonic landscapes in Iceland

Figures 29, 30, and 31 show tectonic landscapes in Iceland.

Figure 29 Rift valley, Thingvellir

Figure 30 Greenhouse cultivation, Hvergerdi

Figure 31 Geothermal energy, Nejsvellir

A Make a sketch diagram of the landform shown in Figure 29. (*4M*)

B Describe the main characteristics of the landscape shown. (*6L, 3M*)

C Figure 31 shows one advantage of tectonic activity in Iceland. State three advantages of tectonic activity. (*6L, 3M*)

D What hazards do volcanoes pose to a country such as Iceland? (*5L, 3M*)

E Suggest reasons why the volcanic hazard in Iceland is not as great as in a country such as the Philippines. (*8L, 5M*)

F With the use of examples outline ways in which the hazards posed by volcanoes can be managed/controlled. (*10L, 7M*)

23. Volcanic activity in St Lucia

Figure 32 shows the effect of volcanic activity in St Lucia.

Figure 32

1 Most of the rocks in this area were laid down around 10 million years ago. The original volcanic landscape has been completely altered by erosion. There are many ridges of high ground, formed by dykes.

2 Mount Gimie is the largest of a group of volcanic cones which was formed about 1.7 million years ago.
3 Several hundred thousand years ago, there was a very large volcano, centred near to the town of Soufrière.
4 About 40,000 years ago, there was a major ignimbrite eruption from this volcano.
5 After the eruption, the centre of the volcano collapsed, leaving a large caldera. The western side of the caldera is missing.
6 Inside the caldera are fifteen intruded viscous lava plugs. The largest of them are the two pitons.
7 This headland is formed of layers of ash and lava laid down around ten million years ago. The steep cliff has been produced by wave erosion.
8 Many of the headlands of this stretch of coast are formed by old lava flows.

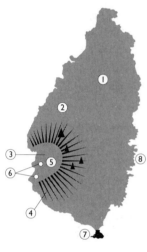

Figure 33 Pitons, St Lucia

Figure 34 shows the sulphur springs at Soufrière.

Figure 34 Sulphur springs, Soufrière

Figures 33 and 34 are volcanic landscapes in St Lucia.

A Define the following terms:
- dyke
- ignimbrite
- lava plug
- fumarole (*8L, 4M*)

B Using the map on page 28 (Figure 36) explain why St Lucia experiences tectonic activity. (*5L, 3M*)

C The Petite Piton is an example of a lava plug. Make a labelled sketch of Figure 33 which shows Petite Piton. On your diagram you should describe the form (shape) of the lava plug. (*4M*)

D Suggest why a volcanic explosion is not likely at Soufrière. (*4L, 3M*)

E Study Figure 32. Suggest why volcanic features have formed
   (i) high ground at 1,
   (ii) steep cliffs at 7, and
   (iii) headlands at 8. (*8L, 6M*)

F Outline the advantages of volcanic activity to areas such as St Lucia. (*10L, 5M*)

24. Features associated with intrusive activity

Figure 35 shows features associated with intrusive activity.

Figure 35

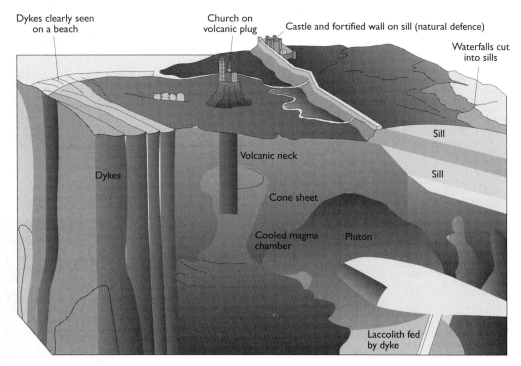

Dykes clearly seen on a beach

Church on volcanic plug

Castle and fortified wall on sill (natural defence)

Waterfalls cut into sills

Sill

Sill

Dykes

Volcanic neck

Cone sheet

Cooled magma chamber

Pluton

Laccolith fed by dyke

A Explain the meaning of the terms 'intrusive' and 'extrusive', as applied to volcanic rocks. (*4L, 2M*)

B State one example of an intrusive rock and one extrusive rock. (*2L, 2M*)

C Describe how the crystal structure of an intrusive rock will differ from that of an extrusive rock. Suggest a reason to explain the difference. (*4L, 3M*)

D What is the difference between a sill and a dyke? (*3L, 2M*)

E Suggest reasons for the settlement at X, but not at Y. (*4L, 3M*)

F Explain the presence of a waterfall at Z. (*5L, 3M*)

G Using examples, show how intrusive volcanic features have affected the human environment. (*7L, 5M*)

H Suggest ways in which human activities have affected intrusive volcanic features. (*6L, 5M*)

25. Geological structure of the Caribbean

Figure 36 shows the geological structure of the Caribbean.

A Draw a labelled cross-section to show what is happening where the South American plate meets the Nazca plate. (*4M*)

B Explain why the Nazca behaves differently from the South American plate. (*3L, 2M*)

C The Eastern Caribbean is an island arc. Describe the main characteristics of an island arc. (*6L, 4M*)

D Suggest reasons how island arcs may be formed. (*7L, 4M*)

E Explain why the Eastern Caribbean experiences volcanic activity and frequent earthquakes. (*6L, 4M*)

F With the use of examples explain how it is possible to predict tectonic hazards. (*7L, 4M*)

G To what extent is it possible to modify the impact of tectonic activity? (*6L, 3M*)

Figure 36

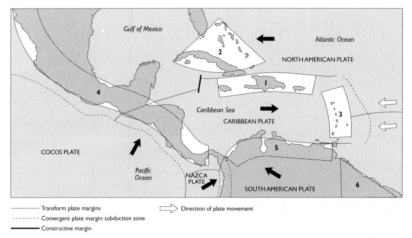

Zone 1 | The oldest rocks in Jamaica, Hispaniola, and Puerto Rico were formed as part of an old island arc system about 100 million years ago. This is no longer a subduction zone. It is now a transform plate margin. There is no volcanic activity here now.

Zone 2 | The Bahamas and Cuba are part of the North American plate. The Bahamas are geologically stable.

Zone 3 | The eastern Caribbean is an *island arc* which follows the line of the subduction zone along the edge of the Caribbean plate. All the islands in the eastern Caribbean are geologically quite recent.

Zone 4 | Mexico and Central America is a very complex area. There is a destructive plate boundary along the Pacific coast. Earthquakes are very common here, and there are many active volcanoes.

Zone 5 | There are more *fold mountains* along the southern edge of the Caribbean, in northern Venezuela and Trinidad.

26. Mass movement

Figure 37 shows some of the causes of mass movement.

A Define the term 'mass movement'. (*3L, 2M*)

B With the use of labelled diagrams, outline the main characteristics of slides and slumps. (*6M*)

C (i) From Figure 37 state four factors that increase the risk of mass movements. (*4L, 2M*)

(ii) For any *two* of these, explain how they increase the risk of mass movement. (*8L, 6M*)

Figure 37

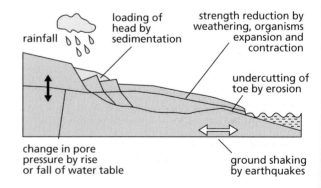

D Using examples, outline how mass movements may prove hazardous to people and their activities. (*8L, 6M*)

E With the use of examples, show how mass movements can be controlled or managed. (*12L, 8M*)

27. Granite scenery

Figure 38 shows a model of tor formation on granite.

Figure 38

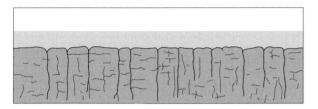

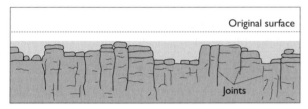

Breakdown of rocks along joints and other lines of weakness

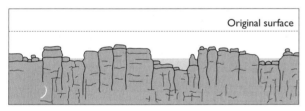

Removal of weathered material to expose tors

A What type of rock is granite? (*1L, 1M*)

B Briefly explain how granite is formed. (*2L, 2M*)

C State *two* characteristics of granite. (*2L, 2M*)

D Explain the processes of freeze thaw and hydrolysis. (*8L, 6M*)

E Describe the main characteristics of a tor and explain why they are described as 'joint controlled'. (*6L, 4M*)

F Outline the ways in which granite is a useful resource for people. (*8L, 5M*)

G Using examples, examine the ways in which human activities have affected granite landscapes. (*8L, 5M*)

28. Permeability and porosity

The table below shows the relative porosity and permeability of selected rock types.

Rock type	Porosity (%)	Relative permeability
Granite	1	1
Basalt	1	1
Shale	18	5
Sandstone	18	500
Limestone	10	30
Clay	45	10
Silt	40	–
Sand	35	1,100
Gravel	25	10,000

A Define the terms 'porosity' and 'permeability'. (*4L, 2M*)

B Study the table and identify

◼ the most porous rock

◼ the least porous rocks

◼ the most permeable rock and

◼ the least permeable rocks. (*4L, 2M*)

C Explain why clay has a relatively high porosity but a low permeability. (*5L, 3M*)

D (i) On the graph paper, Figure 39, plot the data for rock permeability and porosity. The figures for granite, basalt and shale have already been plotted. (*3M*)

(ii) Add a line of best fit to your graph. (*1M*)

(iii) Describe the general relationship between porosity and permeability. (*6L, 3M*)

(iv) Suggest reasons for the relationship described in (iii) (*8L, 4M*)

E For an area you have studied show how rock type has influenced landscape development. (*12L, 7M*)

Figure 39 Graph paper for 28D

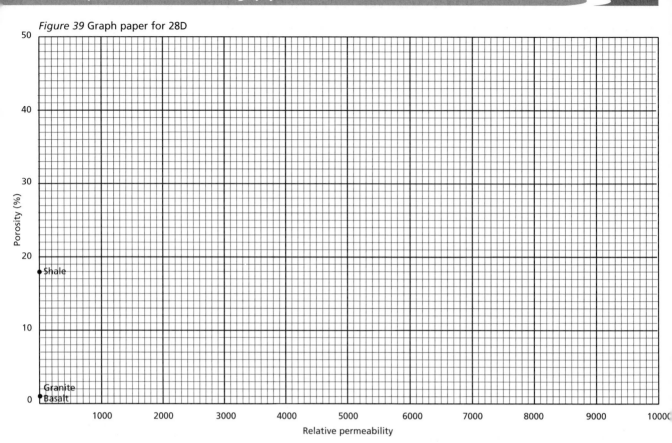

29. Aquifers

Figure 40 shows different types of aquifers.

Figure 40

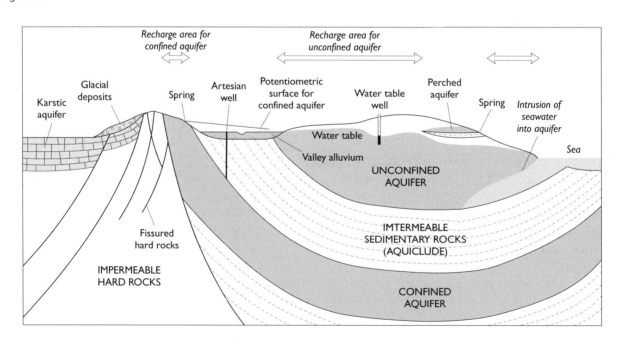

A Define the following terms:

■ groundwater

■ aquifer

■ water table (*6L, 6M*)

B Study the diagram and describe the main characteristics of

■ a perched aquifer

■ an aquiclude

■ a confined aquifer (*9L, 6M*)

C Suggest the likely characteristics of the rocks forming the aquifers. (*3L, 2M*)

D Explain the presence of small quantities of water in the fissured hard rocks. (*3L, 2M*)

E Suggest ways in which groundwater may become contaminated. (*6L, 4M*)

F Outline the social and economic consequences of a decline in groundwater quality. (*8L, 5M*)

30. Geological resources

Figure 41 shows methods of extracting geological resources and some of the main environmental impacts.

Figure 41

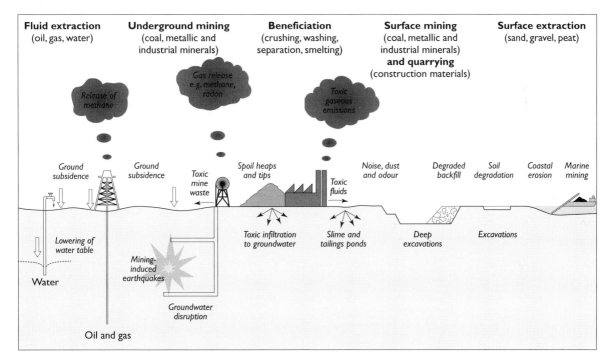

A Identify two liquids extracted from beneath the surface. (*2L, 1M*)

B Briefly describe two problems resulting from the extraction of the named liquids. (*6L, 4M*)

C (i) Suggest a definition of the term *benefication*. (*2L, 1M*)

(ii) Describe the processes that occur during benefication. (*4L, 3M*)

(iii) Explain how benefication may harm the environment. (*8L, 6M*)

D Briefly outline the social and economic advantages of surface mining and quarrying. (*5L, 3M*)

E For a quarry or mine that you have studied:

(i) Name the resource that was being extracted. (*1L, 1M*)

(ii) Outline ways in which the environmental impact was limited or managed. (*8L, 6M*)

31. Construction materials

Figure 42 shows the distribution of selected construction materials in the British Isles.

Figure 42

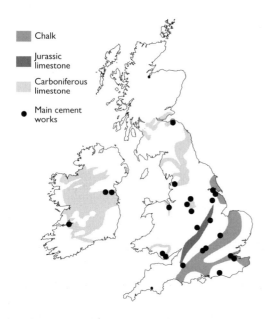

Chalk

Jurassic limestone

Carboniferous limestone

Main cement works

A Describe the distribution of carboniferous limestone in the British Isles. (*4L, 4M*)

Figure 43

B Using an atlas, if necessary, identify the main rock type of
  (i) the Cotswolds and
  (ii) the Chilterns. (*2L, 2M*)

C Comment on the location of the main cement works as shows in Figure 42. (*3L, 3M*)

D Suggest reasons for the location of these main cement works. (*3L, 3M*)

Figure 43 shows the processing and uses of the main geological construction materials.

E What rocks are used to make
  (i) cement, and
  (ii) concrete? (*2L, 2M*)

F Briefly describe the uses of limestone. (*5L, 4M*)

G Suggest a definition for 'armourstone'. (*2L, 1M*)

H Describe the environmental issues associated with mining. (*8L, 6M*)

I With reference to a mine or quarry you have studied, show how it is possible to reduce the impact of mining or quarrying on the environment. (*7L, 5M*)

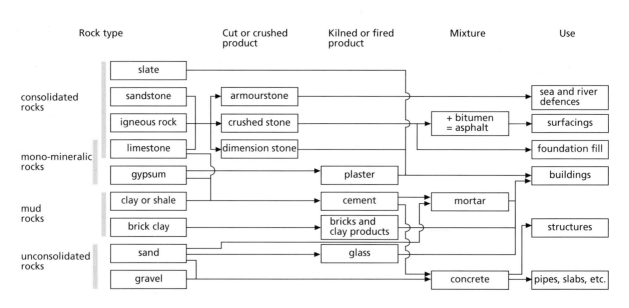

32. Coastal retreat

Figure 44A shows areas of rapid retreat in the UK.

Figure 44A

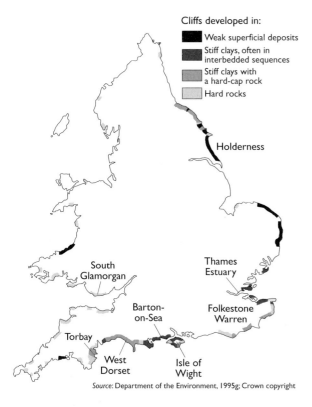

Cliffs developed in:

- ■ Weak superficial deposits
- ■ Stiff clays, often in interbedded sequences
- ▨ Stiff clays with a hard-cap rock
- ▢ Hard rocks

Holderness

South Glamorgan

Thames Estuary

Barton-on-Sea

Folkestone Warren

Torbay

West Dorset

Isle of Wight

Source: Department of the Environment, 1995g; Crown copyright

Figure 44B shows rates of retreat by rock type for selected locations in the UK.

Figure 44B

LOCATION	GEOLOGY	EROSION (metres per century)
Holderness	Glacial drift	120
Cromer	Glacial drift	96
Folkestone	Clay	28
Isle of Thanet	Clay	7–22
Seaford head	Chalk	126
Beachy head	Chalk	106
Barton	Barton beds (clay)	58

A Describe the distribution of rapid retreat in the UK. (*6L, 4M*)

B Which two areas have the highest rates of retreat? (*2L, 2M*)

C Suggest reasons why rates of retreat are expressed in m/century rather than cm/year. (*4L, 3M*)

D Explain *two* reasons why cliff retreat may be rapid. (*6L, 4M*)

E Distinguish between the terms 'hand-' and 'soft-engineering' when applied to coastal management. (*2L, 2M*)

F Suggest why it is important to manage Britain's coastline. (*6L, 4M*)

G For a coastal area you have studied, describe how the coastline is being protected, and comment on the effectiveness of the measures used. (*9L, 6M*)

33. Sea level changes

Figure 45A shows sea level changes around Britain over the last 20,000 years.

Figure 45A

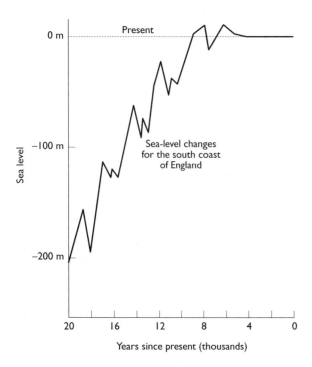

A Briefly describe the meaning of the terms 'eustatic change' and 'isostatic change'. (*6L, 4M*)

B Explain how eustatic changes in sea level may be brought about. (*6L, 4M*)

C Describe how isostatic changes are caused. (*6L, 4M*)

D Compared with current sea levels, by how much did sea levels differ

  (i) 20,000 years ago

  (ii) 5000 years ago? (*2L, 2M*)

Figure 45B shows isostatic changes in Britain since the last glaciation.

Figure 45B

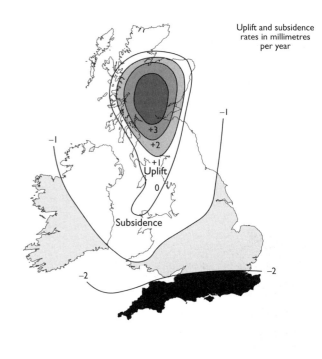

E Which parts of the UK are

  (i) rising most as a result of isostatic change and

  (ii) sinking most rapidly as a result of isostatic change? (*2L, 2M*)

F (i) Name one landform that is associated with a rise in sea level. (*1L, 1M*)

  (ii) Describe this feature, and account for its formation. (*6L, 4M*)

G Briefly suggest ways in which rising sea levels may affect human activity. (*6L, 4M*)

34. Erosional features

Figure 46 shows Durdle Door, in Dorset.

Figure 46

A Identify the features A, B and C. (*3L, 3M*)

B Briefly explain how Feature C may have been formed. (*4L, 3M*)

C Suggest how rock type influences the landscape of D. (*4L, 3M*)

D (i) Describe the nature of the beach shown at B. (*3L, 2M*)

(ii) Suggest two possible sources of beach material at B. (*4L, 2M*)

E (i) Describe the profile of Feature A. (*3L, 2M*)

(ii) Suggest reasons for its profile. (*5L, 3M*)

F (i) Suggest how features A and B may change over time. (*7L, 4M*)

(ii) Suggest reasons why there is no coastal management in this area. (*5L, 3M*)

35. Coastal management

Figure 47 shows a model of coastal management strategies.

A Define the term 'coastal management'. (*2L, 2M*)

B Distinguish between 'hard engineering' and 'soft engineering'. (*2L, 2M*)

C Identify *two* methods of soft engineering as shown on Figure 47. (*2L, 2M*)

Figure 47

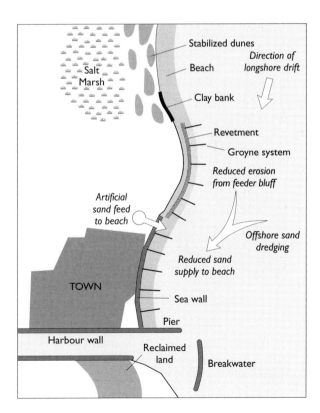

D (i) What is a groyne? (*1L, 1L*)

(ii) Using a sketch diagram, suggest the likely distribution sediment around groynes 50 years after the groyne's construction. Suggest reasons to support your answer. (*6L, 6M*)

E Outline the consequences of *onshore* sand dredging. (*3L, 3M*)

F Suggest guidelines for *offshore* sand dredging to minimise the impact of dredging. (*3L, 2M*)

G (i) What are the benefits of sea walls? (*3L, 2M*)

(ii) Outline some of the disadvantages of using sea walls as a form of coastal management. (*6L, 4M*)

H For an area that you have studied describe the problems encountered, and the remedies that have been used. How successful has the scheme been? (*9L, 6M*)

36. Rias

Figure 48 shows how rias are formed.

A (i) What is a 'ria'? (*2L, 2M*)

(ii) Describe the main characteristics of rias. (*4L, 3M*)

(iii) Suggest how rias are formed. (*6L, 4M*)

B Figure 49 shows the development of rias in St Lucia.

Figure 48

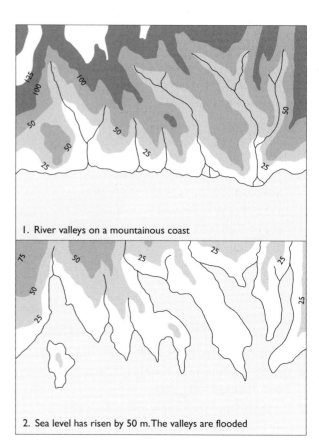

1. River valleys on a mountainous coast

2. Sea level has risen by 50 m. The valleys are flooded

Figure 49

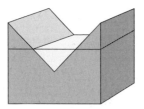
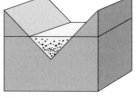
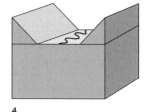

1. 2. 3. 4.

(i) In what ways does the Roseau Valley differ from the typical ria as shown in Figure 48? (*1L, 1M*)

(ii) Describe the cross-section of the Roseau Valley as shown in Stage 4. (*4L, 3M*)

(iii) Explain how the valley achieved this shape. (*4L, 3M*)

C In what ways do rising sea levels pose a threat to modern society? (*6L, 4M*)

D Suggest ways in which it may be possible to counteract or manage rising sea levels. (*8L, 5M*)

37. Rapid coastal erosion

Figure 50A shows a simplified geological map of England and Wales. Figure 50B shows the main characteristics of eroding cliffs in England and Wales.

A Describe the distribution of eroding cliff types in England and Wales. (*5L, 4M*)

B Suggest how rates of erosion may differ between the clays and superficial deposits and the hard rocks. (*3L, 2M*)

C Suggest different reasons for erosion at X and Y. (*3L, 2M*)

D Describe ways in which human activity may lead to an increase in coastal erosion. (*5L, 4M*)

E Outline the physical conditions which may lead to an increase in coastal erosion over the next 50 years. (*5L, 4M*)

F Suggest why there are low rates of erosion at Z_1 and Z_2. (*3L, 2M*)

G For an area you have studied, describe the ways in which the coastline is managed. (*5L, 4M*)

H To what extent has the management of the area achieved its stated aims? (*6L, 5M*)

Figure 50A

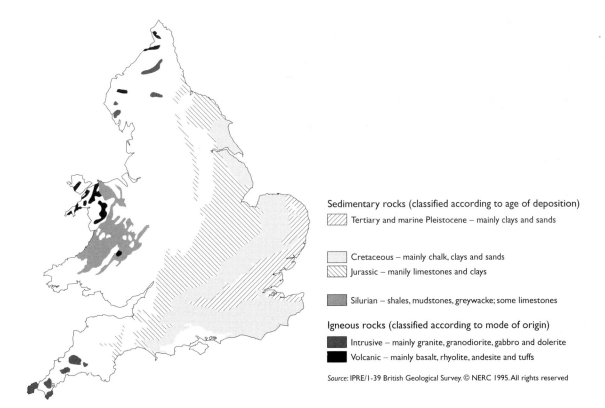

Sedimentary rocks (classified according to age of deposition)

///// Tertiary and marine Pleistocene – mainly clays and sands

☐ Cretaceous – mainly chalk, clays and sands

\\\\\ Jurassic – manily limestones and clays

▨ Silurian – shales, mudstones, greywacke; some limestones

Igneous rocks (classified according to mode of origin)

▮ Intrusive – mainly granite, granodiorite, gabbro and dolerite
■ Volcanic – mainly basalt, rhyolite, andesite and tuffs

Source: IPRE/1-39 British Geological Survey. © NERC 1995. All rights reserved

Figure 50B

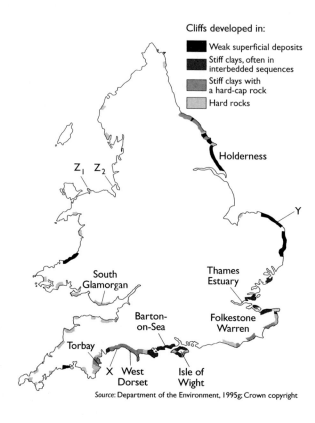

Cliffs developed in:

▪ Weak superficial deposits

▪ Stiff clays, often in interbedded sequences

▪ Stiff clays with a hard-cap rock

▫ Hard rocks

Holderness

Z_1 Z_2

Y

South Glamorgan

Thames Estuary

Barton-on-Sea

Folkestone Warren

Torbay

X West Dorset

Isle of Wight

Source: Department of the Environment, 1995g; Crown copyright

38. CO$_2$ levels and ozone levels

Figure 51 shows changes in CO$_2$ levels and CFCs in the twentieth century.

Figure 51

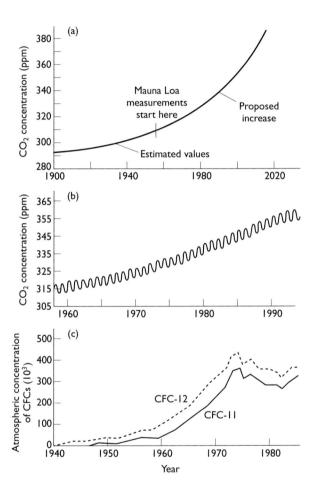

A (i) Describe the changes in CO$_2$ as shown in Figure 51. (*4L, 3M*)

(ii) Suggest why the annual pattern shown in Figure 51 shows an increase in the concentration of CO$_2$ in winter. (*3L, 2M*)

(iii) Describe and suggest reasons for changes in the curve for CFCs as shown in Figure 51C. (*5L, 4M*)

B (i) Define the term 'greenhouse gas'. (*1L, 1M*)

(ii) What is meant by the term 'greenhouse effect'? (*5L, 3M*)

(iii) Suggest ways in which human activity has led to an increase in the amount of greenhouse gases in the atmosphere. (*5L, 3M*)

C Outline some of the possible environmental impacts of an increased greenhouse effect. (*5L, 4M*)

D Briefly describe why it is proving difficult to reduce emissions of greenhouse gases. (*7L, 5M*)

39. Air pressure

Figure 52 shows mean global pressure in winter and summer.

A (i) Describe the pattern of mean air pressure in the northern hemisphere in winter. (*5L, 3M*)

(ii) How does this compare with the pattern of air pressure in summer? (*5L, 3M*)

Figure 52

(a)

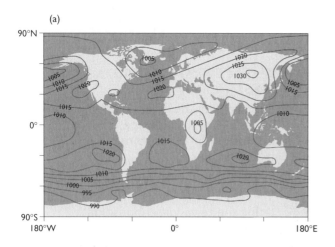

(b)

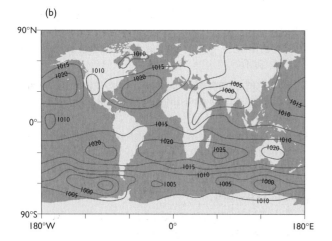

Figure 53

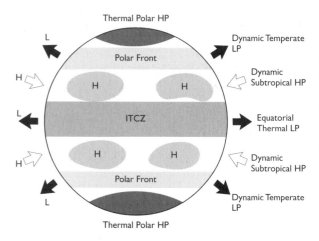

E On a copy of Figure 53, complete the diagram by adding main global winds (polar easterlies, westerlies, and trades). (*6*)

F Briefly explain two reasons for the air movement you have drawn. (*8L, 4M*)

40. Sea breezes

Figure 54A shows the arrival times for sea breezes, and the weather associated with sea breezes.

A What is meant by the term 'sea breeze'? (*2L, 2M*)

B With the use of diagrams briefly explain how land-sea breezes are formed. (*10L, 6M*)

C (i) State the average arrival time of the sea breeze at Reading. (*1L, 1M*)

(ii) Calculate the speed of the sea breeze moving from Thorney Island to Reading. (*3L, 2M*)

Figure 53 shows a simplified sketch map of high and low pressure belts over the globe.

B (i) Explain why there is a zone of high pressure in the subtropics. (*5L, 3M*)

(ii) Why is there low pressure at the equator? (*4L, 2M*)

C State what is happening to air at (i) low pressure zones and (ii) high pressure zones. (*4L, 2M*)

D What is happening at the polar front? (*4L, 2M*)

Figure 54A

Figure 54B

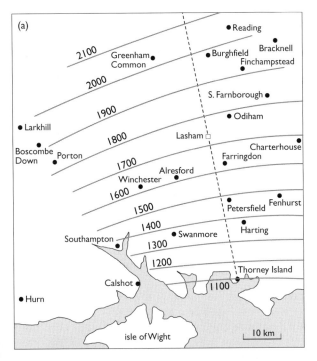

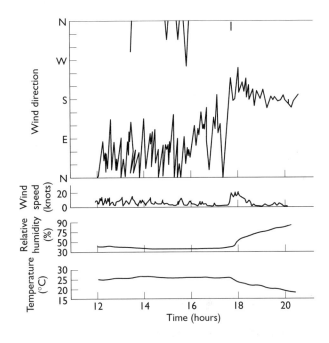

Figure 54B shows the effect of the sea breeze which arrived at Porton Down around 6pm on 24 July 1959.

D (i) Describe the changes in the windspeed as the sea breeze arrived. (*3L, 3M*)

(ii) How did relative humidity change as the sea breeze passes over? (*3L, 2M*)

(iii) Suggest reasons for the changes noted in D(i) and (ii). (*4L, 3M*)

(iv) Describe and explain how temperatures changed following the arrival of the sea breeze. (*9L, 6M*)

41. Air masses

Figure 55A shows some of the main air mass sources over Europe.

A Define the term 'air mass'. (*4L, 2M*)

B Suggest the main characteristics of
(i) polar maritime air, and
(ii) tropical continental air. (*4L, 4M*)

C Suggest what happens to
(i) polar continental air in winter as it passes over the North Sea,
(ii) tropical maritime air in summer as it passes over the cold Canary current. (*4L, 4M*)

Figure 55A

Figure 55B

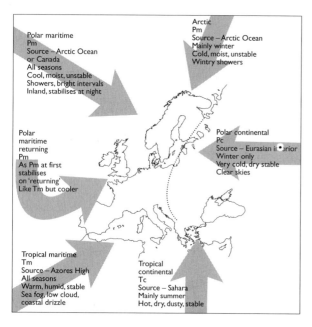

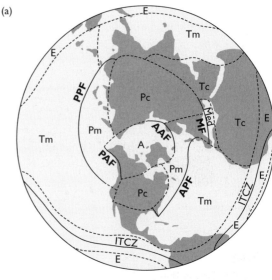

(a)

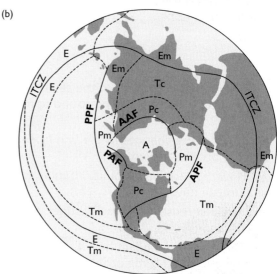

(b)

Figure 55B shows air mass source regions and frontal zones in winter and summer.

D (i) What is the (Atlantic) Polar Front? (*1L, 1M*)

(ii) Describe the annual variation in the location of the Atlantic Polar Front in relation to the British Isles. (*4L, 2M*)

(iii) What are the implications of the location of the polar front for the weather experienced in the British Isles? (*6L, 4M*)

E Using examples, examine the hazards that different air masses bring to the British Isles. (*12L, 8M*)

42. Temperature inversions

Figure 56 shows some of the conditions that promote temperature inversions and local air pollution.

A Define the term 'temperature inversion'. (*2L, 2M*)

B Describe the changes in temperature as shown in the line graph. (*3L, 3M*)

C Outline the synoptic (weather) conditions that give rise to temperature inversions. (*3L, 3M*)

D Describe the main geographic locations in which temperature inversions are likely to occur. (*4L, 4M*)

PPF Pacific Polar Front   APF Atlantic Polar Front
PAF Pacific Arctic Front   AAF Atlantic Arctic Front
MF   Mediterranean Front

E (i) Describe the problems associated with temperature inversions. (*6L, 4M*)

(ii) Outline the factors that increase the impact of these problems. (*9L, 5M*)

F Suggest how, and why, the impact of temperature inversions may differ between LEDCs and MEDCs. Use examples to support your answer. (*8L, 4M*)

Figure 56

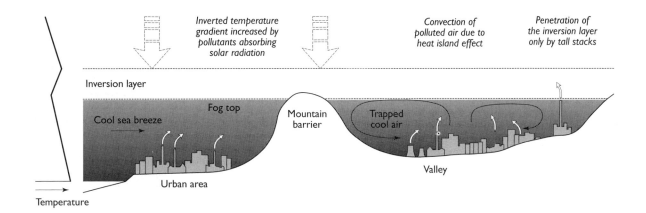

43. Smog

Figure 57 shows the build up of air pollutants in winter conditions.

Figure 57

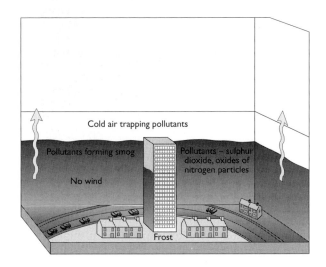

A What is 'smog'? (*3L, 2M*)

B Describe the conditions that lead to the formation of smog. (*6L, 4M*)

C Describe the characteristics of a temperature inversion. (*3L, 2M*)

D Briefly explain how temperature inversions form. (*6L, 4M*)

E Suggest why smogs are more common in winter. (*3L, 2M*)

F Briefly explain why smogs are more likely during calm, anticyclonic conditions. (*4L, 3M*)

G (i) Suggest ways in which air quality in urban areas can be improved.

(ii) For an area that you have studied, comment on the success of schemes to improve air quality. (*10L, 8M*)

44. Temperature patterns

Figure 58 shows mean January and July temperatures in the UK.

A (i) Describe the pattern of temperatures in January. (*4L, 4M*)

(ii) How do you account for this pattern? (*4L, 4M*)

B (i) What is the annual range in temperature at

■ Anglesey (A)?

■ London (B)?

■ Land's End (C)? (*3L, 3M*)

(ii) Suggest reasons for variations in annual temperature patterns. (*4L, 3M*)

C Outline the natural factors that affect the temperature of a location (*9L, 5M*)

D With the use of examples explain how human factors affect the temperature of an area. (*11L, 6M*)

Figure 58

January

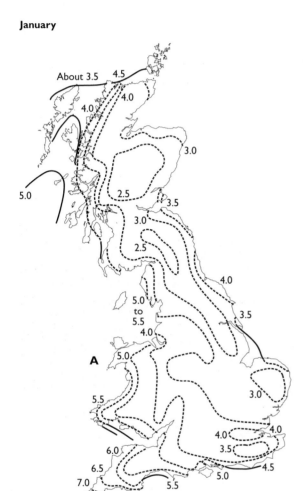

July

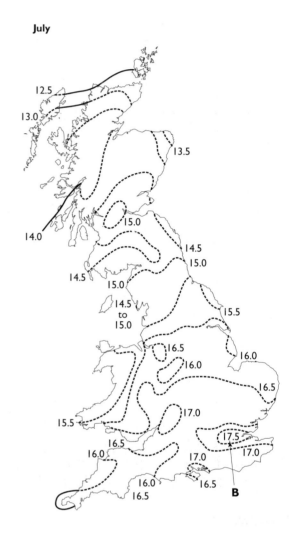

Temperature (°C) corrected to mean sea level

Source: Met. Office

Figure 59

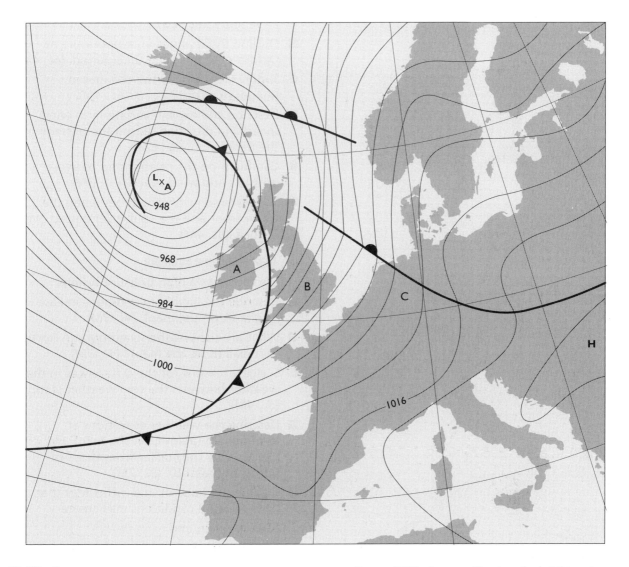

45. Weather maps

A What is the pressure (in Mb) of centre A on the weather chart, Figure 59? (*1L, 1M*)

B Name the type of weather system shown in Figure 59. (*1L, 1M*)

C Identify the type of front located over
(i) the North Sea, and
(ii) the west of Scotland. (*2L, 2M*)

D The weather at A, B and C was as follows:

A   4°C, westerly wind, showers, speed 24 knots, 7/8 cloud cover.

B   12°C, dry, southerly wind, 18 knots, 5/8 cloud cover.

C   6°C, south-east wind, drizzle, 12 knots, 6/8 cloud cover.

Draw four weather station maps to show the weather conditions at each of the three locations. (*16L, 12M*)

E Write a weather forecast for Denmark suggesting how and why the weather will change over the next 24 hours. (Assume that the whole system will pass over during the 24-hour period.) (*15L, 9M*)

46. Low pressure over Oxford

Figure 60 shows the passage of a depression over Oxford and the accompanying weather.

Figure 60

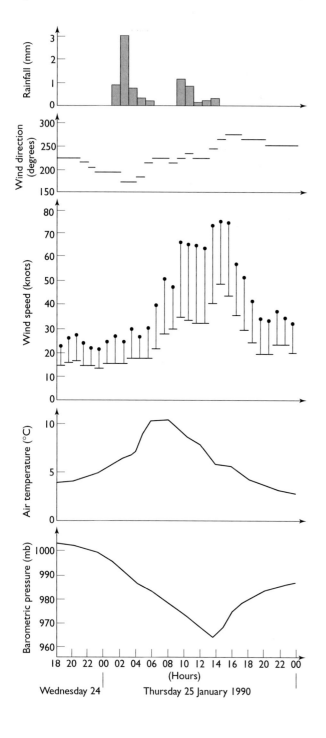

Wednesday 24   Thursday 25 January 1990

A Describe the changes in rainfall between 1800 hours on 24 January and 2400 hours on 25 January. (*6L, 4M*)

B Outline the changes in temperature at Oxford over the same time period. (*6L, 4M*)

C Suggest reasons for the changes in temperature noted in B. (*6L, 4M*)

D Describe and explain variations in wind speed and direction during the passage of the depression. (*8L, 6M*)

E Comment on the relationship between air pressure and wind speed. (*3L, 3M*)

F Suggest and justify the times when the warm front and the cold front passed over Oxford. (*6L, 4M*)

47. High pressure systems

Figure 61 shows a high pressure system.

A Describe the weather conditions in East Anglia (A on the map). (*4L, 3M*)

B What are the weather conditions in North Cornwall (B on the map)? (*4L, 3M*)

C Suggest reasons for the differences in the weather between the two weather stations. (*6L, 4M*)

D Describe the weather conditions at Birmingham (point C on the map) and explain why they differ from those of Cornwall (point B). (*9L, 7M*)

E Outline the hazards and difficulties that high pressure conditions may create

(i) in summer, and

(ii) in winter. (*12L, 8M*)

Figure 61

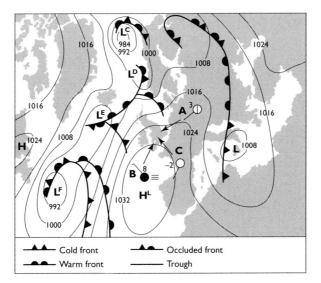

48. Soil and land degradation

Figure 62 shows the amount and causes of degraded land by continent.

Figure 62

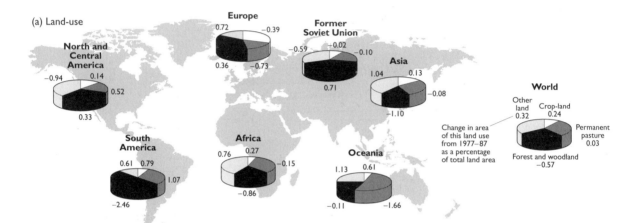

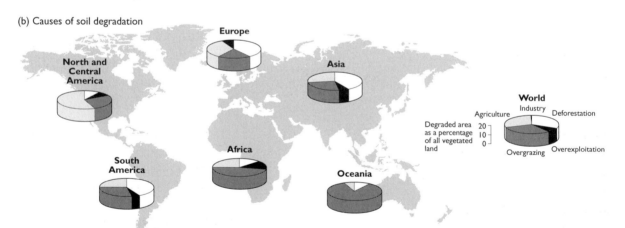

A Suggest a definition for the term 'degradation'. (*2L, 2M*)

B What are the likely characteristics of land that is degraded? (*2L, 2M*)

C In which continent is there most degradation? (*1L, 1M*)

D In which continent are (i) agriculture and (ii) overgrazing the main causes of degradation? (*2L, 2M*)

E Explain how overgrazing may lead to soil degradation. (*6L, 4M*)

F In which continent is industry a major cause of soil degradation? (*1L, 1M*)

G Outline the social and economic problems associated with soil degradation. (*12L, 8M*)

H Suggest ways in which it may be possible to restore degraded land. (*9L, 5M*)

49. Deciduous woodland ecosystems

Figure 63 shows a food web in temperate deciduous forest.

A Explain what is meant by the term 'food web'. (*2L, 2M*)

B What is meant by the term 'trophic level'? (*2L, 2M*)

C Identify four trophic levels in the diagram below. (*2L, 2M*)

D (i) Explain how energy from the sun is converted to food energy then passed through the food web. (*4L, 3M*)

(ii) Suggest why less energy is available at each successive trophic level. (*4L, 3M*)

E What is meant by the term 'biome'? (*3L, 2M*)

F Describe the main characteristics of a temperate deciduous forest ecosystem or of any other forest ecosystem that you have studied. (*8L, 5M*)

G Using examples, show how human activities are affecting forest ecosystems. (*10L, 6M*)

Figure 63

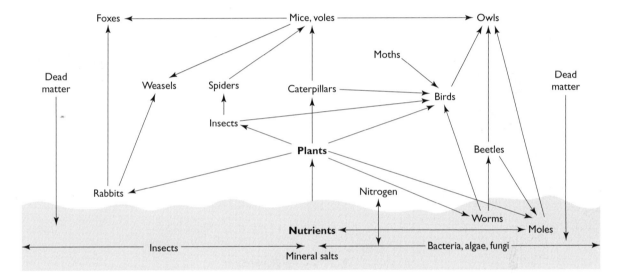

50. Nutrient cycles

Figure 64A shows a model of a nutrient cycle.

Figure 64B shows the nutrient cycle for a coniferous forest.

Figure 64C shows the result of cultivation on a rainforest nutrient cycle.

A Name two sources of nutrients shown in Figure 64A. (*2L, 2M*)

B Define the terms 'litter' and 'biomass'. (*2L, 2M*)

C Explain the differences between leaching and run-off. (*2L, 2M*)

D Explain why

(i) the litter store is the largest store in this cycle, and

(ii) the input from weathering is limited. (*6L, 4M*)

E Draw a labelled diagram to show the nutrient cycle for a tropical rainforest ecosystem. (*5M, 8L*)

F Why does it differ from that of the coniferous forest? (*7L, 5M*)

G Comment on the likely impact of cultivation on the nutrient cycle of the rainforest. (*8L, 5M*)

Figure 64A

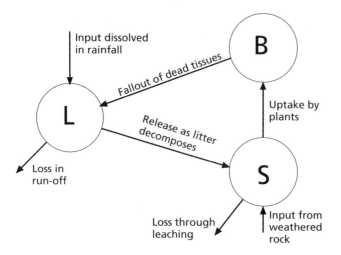

KEY
B - Biomass
S - Soil
L - Litter

The size of the nutrient stores (B,L,S) is proportional to the quantity of nutrients stored. The thickness of the arrows indicates the amount of nutrients transferred.

Figure 64B

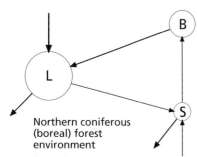

Northern coniferous (boreal) forest environment

Figure 64C

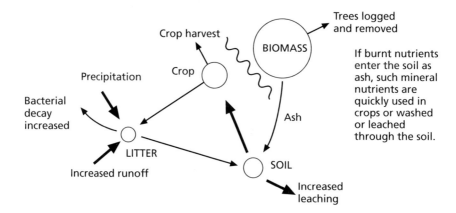

Trees logged and removed

If burnt nutrients enter the soil as ash, such mineral nutrients are quickly used in crops or washed or leached through the soil.

51. Succession

Figure 65 is a model of succession.

A Define the term 'succession'. (*2L, 2M*)

B Explain what is meant by the following terms:

(i) pioneer species,

(ii) climatic climax vegetation. (*2L, 2M*)

C What is the climatic climax vegetation for Southern England? (*1L, 1M*)

D Describe and explain the changes in biomass with succession as shown in Figure 65. (*4L, 3M*)

E How and why does nutrient availability change during succession? (*6L, 4M*)

F With reference to a named ecosystem, outline the changes in species composition as succession takes place. (*10L, 6M*)

G In what ways have human activities affected the succession of your chosen ecosystem. (*8L, 5M*)

Figure 65

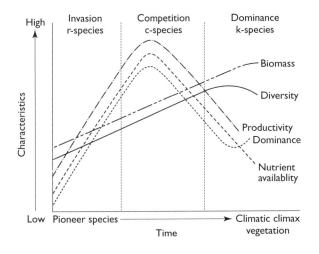

52. Hydoseres

Figure 66A shows a model of succession in fresh water – a hydrosere.

Figure 66A

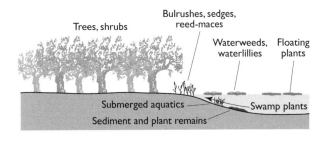

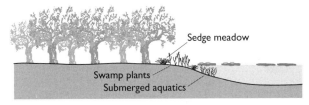

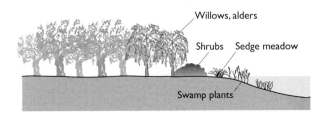

Figure 66B shows a theoretical model of human interference.

Figure 66B

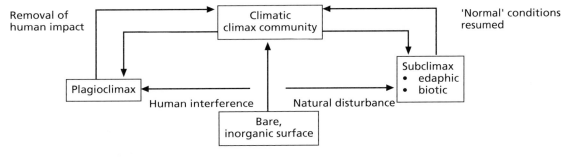

A Name two plants associated with fresh water, as shown in Figure 66A. (*2L, 1M*)

B Name two tree species that are commonly found close to fresh water in Figure 66A. (*1L, 1M*)

C Explain why the lake decreases in size over time. (*6L, 4M*)

D Explain the process of succession as shown in Figure 66B. (*7L, 5M*)

E Outline the main differences between a subclimax and a plagioclimax, as shown in Figure 66B. (*8L, 6M*)

F With reference to a named hydrosere, identify the ways in which human activities have affected hydroseres. (*11L, 8M*)

53. Island biogeography

Figure 67 shows some general design rules based on the theory of island biogeography. The designs on the left ('good') should ensure a better chance of survival than the designs on the right ('not so good').

Figure 67

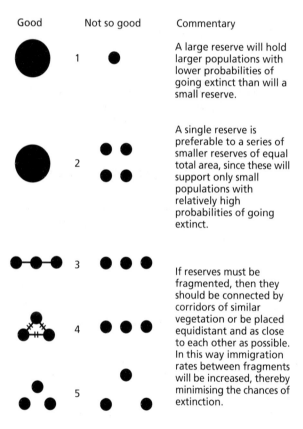

A Why are large reserves better for conservation than small reserves? (*2L, 2M*)

B Why is a single reserve better than a large number of smaller reserves adding up to the same area? (*2L, 2M*)

C Why are fragmented reserves that are linked by a corridor better than those that are not linked? (*3L, 2M*)

D Why are reserves that are close together better than those that are far apart? (*4L, 3M*)

E With reference to a nature reserve or conservation area that you have studied

(i) explain why it was designated as a reserve or conservation area (*3L, 2M*)

(ii) outline the methods used to achieve its aims (*9L, 6M*)

(iii) evaluate the success of the management strategy designed to protect your chosen area. (*12L, 8M*)

54. Savanna ecosystems

Figure 68 shows the climate associated with a savanna grassland.

A Name the climate type shown in Figure 68. (*1L, 1M*)

Figure 68

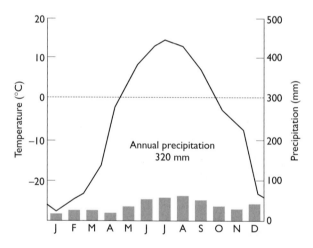

B Describe the climate's main features. (*7L, 5M*)

C In what ways have the vegetation and animals in savanna grasslands adapted to the environment in which they live? (*6L, 4M*)

D Describe three features of soils associated with savanna grasslands. (*5L, 3M*)

E Explain how human activities have affected savanna ecosystems. (*7L, 5M*)

F To what extent is it possible to reduce the impact that human activities are having on savanna ecosystems? (*9L, 7M*)

55. UK soils

Figure 69 shows the main soil types found in the UK.

Figure 69

Key

Shallow rocky and peaty soil

Podzols (with brown soils)

Brown soils (with gley soils)

Gley soils (with brown soils)

Argillic brown soils and gley soils

Alluvial gley soils

Urban soils

Rendzinas

A (i) Describe the distribution of rendzina soils. (*3L, 2M*)

(ii) Describe the main characteristics of a rendzina soil. (*4L, 4M*)

(iii) With what rock types are rendzinas associated? (*1L, 1M*)

B (i) What is a gleyed soil? (*3L, 2M*)

(ii) Describe the main characteristics of a gleyed soil. (*4L, 3M*)

(iii) Explain why gleyed soils can be found (a) in highland areas and (b) in lowland areas. (*9L, 6M*)

C Using examples, outline the ways in which human activities affect soils. (*11L, 7M*)

56. Soil structure

Figure 70 shows soil structure.

A (i) State the composition of the soils A, B, C and D. (*4L, 4M*)

(ii) Plot the following soil on the graph and name its type.

	Sand	Silt	Clay
E	75	10	15
F	10	40	50

(*2L, 2M*)

B Several of the soils are described as 'loams'. Explain what a loam is, and why it is good for agriculture. (*5L, 4M*)

C Distinguish between the terms 'soil structure' and 'soil texture'. (*4L, 2M*)

D Explain why soil structure and soil texture are important to farmers. (*5L, 4M*)

E Suggest how structure and texture can be altered by human activities. (*5L, 4M*)

F Outline the ways in which agricultural activities may alter soils. (*10L, 5M*)

Figure 70

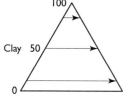

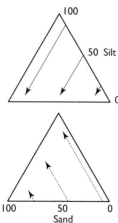

Particle type	Diameter (mm)
Clay	<0.002
Silt	<0.02
Sand	<0.2
Gravel	<2
Fine	>2

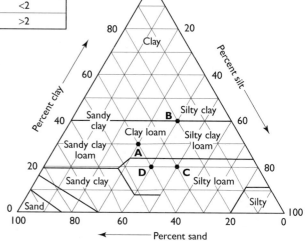

57. Intrazonal soils

Figure 71 shows a model of soil types and geology in southern England.

A (i) Suggest the type of soil likely to be found on

■ sands and gravel

■ limestone. (*2L, 2M*)

(ii) Identify the soils in Figures 71B and 71C.

B In what ways might

(i) chalk and

(ii) clay influence soil development? (*4L, 4M*)

C (i) Describe the main characteristics of a brown earth. (*2L, 2M*)

(ii) Draw a labelled soil profile for a brown earth. (*9L, 4M*)

(iii) How might a brown earth be modified in location X? (*2L, 2M*)

D (i) What type of soil would you expect to find at Y? (*1L, 1M*)

(ii) Explain why the soil at Y is likely to be quite shallow. (*3L, 3M*)

(iii) Describe how the soil at Y might differ from a soil at Z. (*4L, 2M*)

E With the use of examples describe the ways in which human activities affect soil development. (*8L, 5M*)

Figure 71A

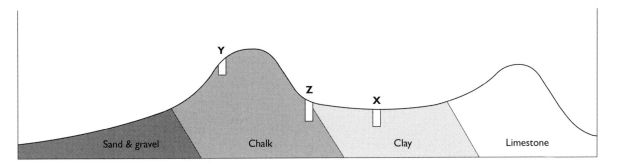

Figure 71B

Figure 71C

58. Soil erosion

Figure 72A shows some of the causes of soil erosion while Figure 72B shows methods to control soil erosion.

A Describe the poor farming techniques and methods of land use as shown in 1, 2, 3 and 9. (*4L, 4M*)

B What impact could soil erosion have on farmland (5,7), settlement (5), water resources (6), and communications (4,8)? (*12L, 8M*)

C Describe the methods of soil conservation that are taking place at 1, 2 and 3. (*9L, 6M*)

D Suggest some of the benefits of soil conservation at 6, 7, 8 and 9. (*10L, 7M*)

Figure 72A & B

(a)

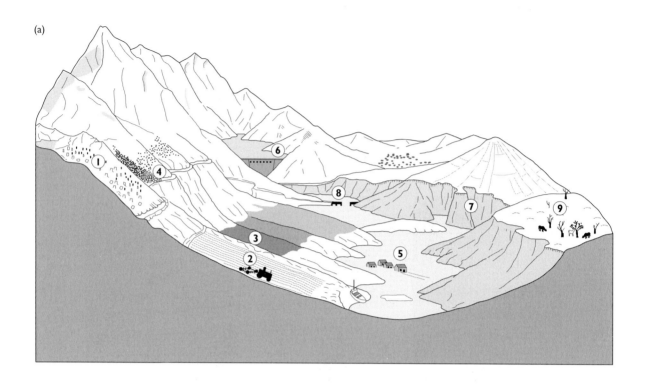

(b)

59. Protected landscapes

Figure 73 shows some of the protected areas of England and Wales.

Figure 73

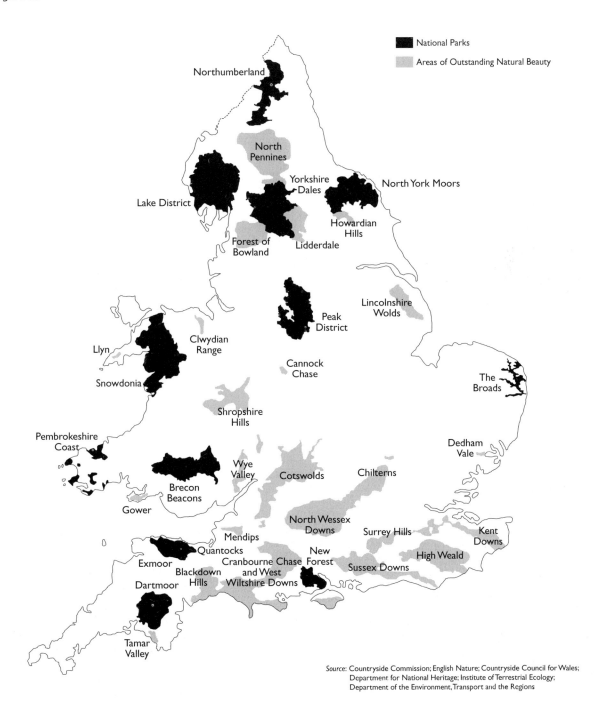

National Parks

Areas of Outstanding Natural Beauty

Northumberland

North Pennines

Yorkshire Dales

North York Moors

Lake District

Howardian Hills

Forest of Bowland

Lidderdale

Peak District

Lincolnshire Wolds

Llyn

Clwydian Range

Cannock Chase

The Broads

Snowdonia

Shropshire Hills

Pembrokeshire Coast

Dedham Vale

Wye Valley

Cotswolds

Chilterns

Brecon Beacons

Gower

North Wessex Downs

Surrey Hills

Kent Downs

Mendips

High Weald

Exmoor

Quantocks

New Forest

Sussex Downs

Cranbourne Chase and West Wiltshire Downs

Blackdown Hills

Dartmoor

Tamar Valley

Source: Countryside Commission; English Nature; Countryside Council for Wales; Department for National Heritage; Institute of Terrestrial Ecology; Department of the Environment, Transport and the Regions

A What is a National Park? (*2L, 2M*)

B Describe the distribution of National Parks in England and Wales. (*6L, 4M*)

C Outline the ways in which National Parks are managed. (*6L, 4M*)

D For a National Park you have studied, comment on the extent to which it has achieved its aims. (*6L, 4M*)

E What is an Area of Outstanding Natural Beauty (AONB)? (*1L, 1M*)

F Outline the pressures on AONBs such as the High Weald or the Chilterns. (*4L, 3M*)

G How are these pressures likely to change over the next 20 years or so? (*4L, 3M*)

H Suggest ways in which AONBs can be protected. (*6L, 4M*)

60. Environmental degradation – the Aral Sea

Figure 74 shows changes in the size and characteristics of the Aral Sea between 1960 and 1989, and predicted contraction up until 2000.

A Describe the changes in the size of the Aral Sea between 1960 and 2000. (*5L, 3M*)

B Using Figure 75, suggest reasons why the Aral Sea has declined in size. (*4L, 3M*)

C Compare and contrast the change in the volume of water in the Aral Sea and the mineral content of the sea. (*5L, 4M*)

D Explain why the mineral content increases over time. (*3L, 2M*)

E Under what environmental conditions does irrigation occur? (*3L, 2M*)

F Briefly describe *two* contrasting methods of irrigation. (*6L, 4M*)

G Outline the advantages and disadvantages of irrigation. Use examples and case studies to support your answer. (*9L, 7M*)

Figure 74

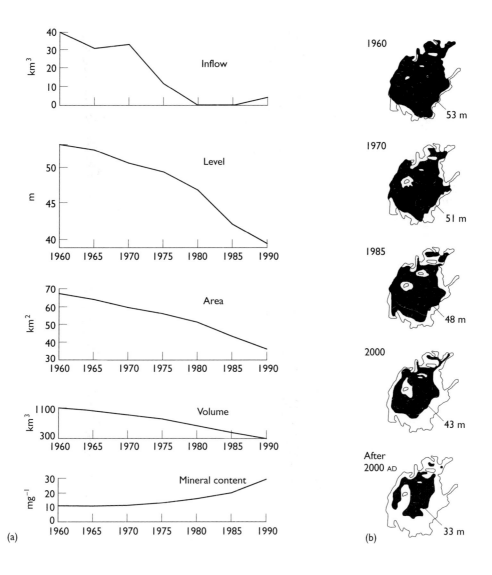

Figure 75

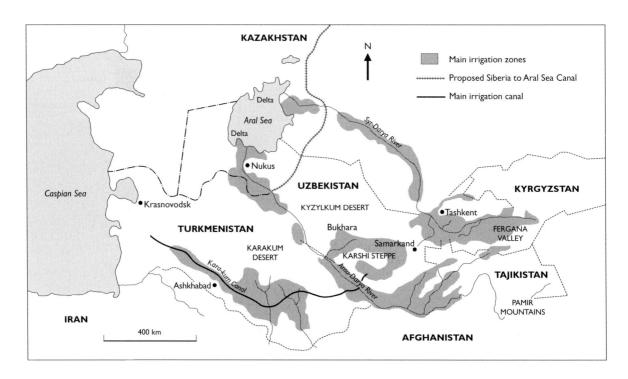

61. World population growth

Figure 76 World Population growth rate 1950–2050

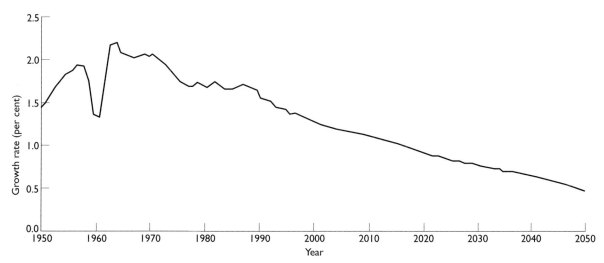

Source: US Census Bureau, International Data Base.

Figure 77 Annual world population change 1950–2050

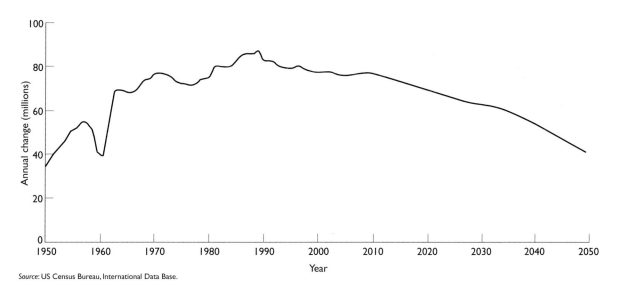

Source: US Census Bureau, International Data Base.

A What is meant by the terms 'growth rate' in Figure 76 and 'annual change' in Figure 77? (2 × [2L, 1M])

B State the growth rate and the annual change for each of the following years: (i) 1960 (ii) 1980 (iii) 2000. (3 × [1L, 1M])

C For Figure 76 suggest one reason for each of the following:

(i) the increase in the growth rate in the early and mid-1950s

(ii) the rapid decline in growth rate in the late 1950s

(iii) the steady fall in growth rate, and predicted fall to 2050, from the early 1960s. (3 × [*3L, 3M*])

D Explain why the peak in annual world population change is about 25 years later than the peak in world population growth rate. (*7L, 4M*)

E Discuss the factors which could cause the predictions for the period 2000–2050 to prove inaccurate. (*12L, 7M*)

62. Population growth in LEDCs and MEDCs

Figure 78 World population growth 1750–2050

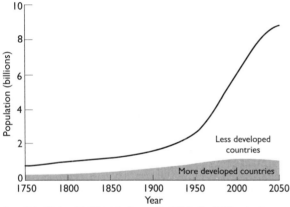

Source: United Nations, *World Population Prospects: The 1998 Revision* (1998); and estimates by the Population Reference Bureau.

A Describe the growth in total world population from 1750 to 2000 and the projected growth to 2050. (*5L, 3M*)

B How has population growth differed between the LEDCs and the MEDCs? (*5L, 3M*)

C Explain the rapid increase in growth in LEDCs from about 1950. (*7L, 5M*)

D Figure 78 shows that population in the MEDCs will begin to decline in the coming decade. Suggest reasons for this. (*7L, 6M*)

E What could happen in order for the United Nations to revise the population projection to 2050? (*11L, 8M*)

63. The model of demographic transition

A Complete the line in Figure 79 showing the change in total population (*3M*)

B State two reasons for the fluctuations in the death rate in Stage 1. (2 × [*3L, 2M*])

C State two reasons for the considerable decline in death rate in Stages 2 and 3. (2 × [*3L, 2M*])

D Suggest why the fall in the birth rate lagged significantly behind the fall in the death rate. (*7L, 4M*)

E What are the likely causes of variations in the birth rate in Stage 4? (*8L, 5M*)

F The evidence of a fifth stage is beginning to appear in some countries. Describe and explain the expected characteristics of this stage. (*8L, 5M*)

Figure 79 The demographic transition model

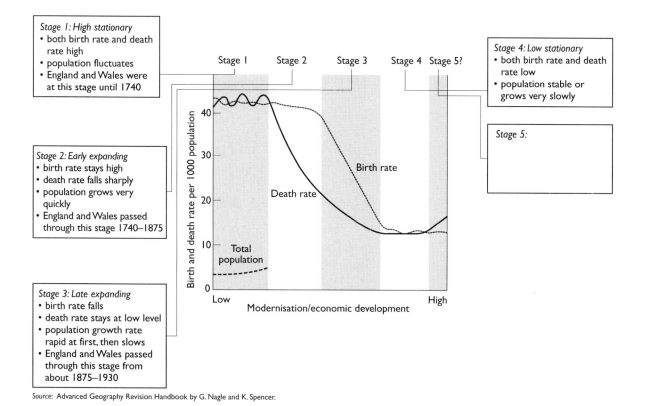

Stage 1: High stationary
• both birth rate and death rate high
• population fluctuates
• England and Wales were at this stage until 1740

Stage 4: Low stationary
• both birth rate and death rate low
• population stable or grows very slowly

Stage 5:

Stage 2: Early expanding
• birth rate stays high
• death rate falls sharply
• population grows very quickly
• England and Wales passed through this stage 1740–1875

Stage 3: Late expanding
• birth rate falls
• death rate stays at low level
• population growth rate rapid at first, then slows
• England and Wales passed through this stage from about 1875–1930

Source: Advanced Geography Revision Handbook by G. Nagle and K. Spencer.

64. Demographic transition in England and Wales

Figure 80 shows changes in birth and death rates between 1700 and 2000.

A Identify and explain two factors which caused the death rate to rise at particular times. (2 × [*3L, 2M*])

B Identify and explain two factors which caused the death rate to fall sharply at particular times. (2 × [*3L, 2M*])

C In terms of the birth rate suggest reasons for:

(i) the fall in the birth rate in Stage 3,
(ii) the post-war baby boom.
(2 × [*3L, 2M*])

D Describe and explain the variations in natural change from 1700 to 2000. (*8L, 6M*)

E What is the evidence on the diagram that the reliability of demographic data improved during the nineteenth century? (*9L, 7M*)

Figure 80 The UK: changes in birth and death rates 1700–2000

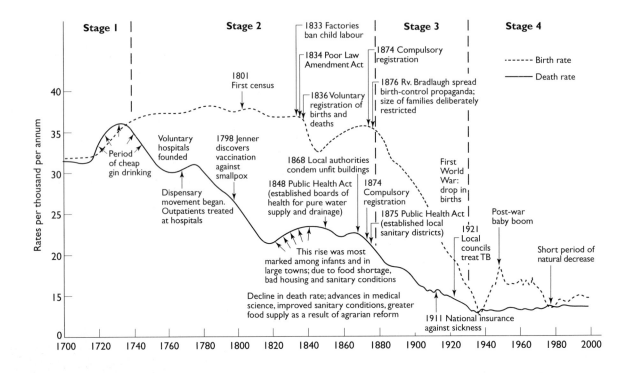

65. Changing population density and distribution in Brazil

Figure 81 shows population in Brazil.

A Define the terms 'population density' and 'population distribution'. (2 × (*2L, 2M*))

B Describe and suggest reasons for the changes in population density in Brazil between 1960 and 1995. (*9L, 6M*)

C Discuss the possible advantages and disadvantages of the low population density of the Amazon basin (Figure 82) and the high population density of the Southeast (Figure 83). (*10L, 7M*)

D What policies might a government pursue to try to influence the distribution of population? (*12L, 8M*)

Figure 81

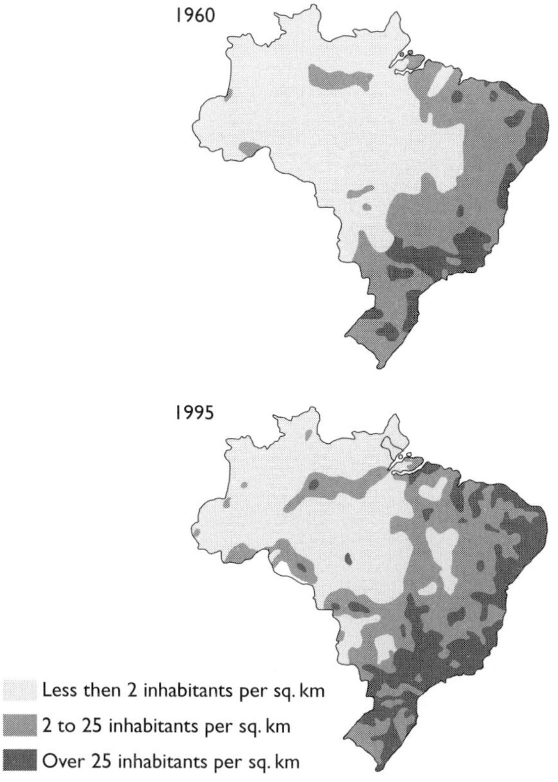

1960

1995

Less then 2 inhabitants per sq. km

2 to 25 inhabitants per sq. km

Over 25 inhabitants per sq. km

Figure 82 The Amazon basin

Figure 83 São Paulo in Southeast Brazil

66. Contraceptive use and fertility in developing countries

A Define the terms 'birth rate' and 'total fertility rate'. (2 × [2L, 2M])

B Why is the birth rate considered to be only a broad indicator of fertility levels? (6L, 4M)

C To what extent does Figure 84 show that fertility varies within both the developed and developing worlds? (6L, 4M)

D Describe and account for the relationship illustrated by Figure 85. (7L, 5M)

E Discuss the full range of factors that can affect fertility levels. (12L, 8M)

Figure 84 Fertility and contraception

	Birth rate	Total fertility rate	% of married* women using contraception (all methods)
World	23	2.9	58
More developed	11	1.5	72
Less developed	26	3.2	55
Africa	39	5.4	24
North America	14	2.0	77
Latin America and Caribbean	24	2.9	68
Asia	23	2.8	60
Europe	10	1.4	71
Oceania	18	2.4	59

Source: *1999 World Population Data Sheet* (For developed countries nearly all data refers to 1997 or 1998 and for less developed countries to some point in the mid to late 1990s.)
* married or 'in union'

Figure 85 Contraceptive use and fertility in developing countries

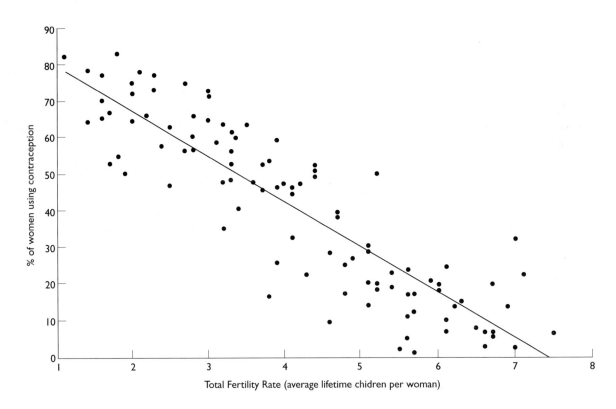

67. The economic effects of reducing fertility

A Define 'fertility' and 'child dependency'. (2 × [*2L, 2M*])

B Explain two factors which could result in a decline in fertility in a country. (*6L, 5M*)

Figure 86 Economic effects of reducing fertility: the Coale-Hoover hypothesis

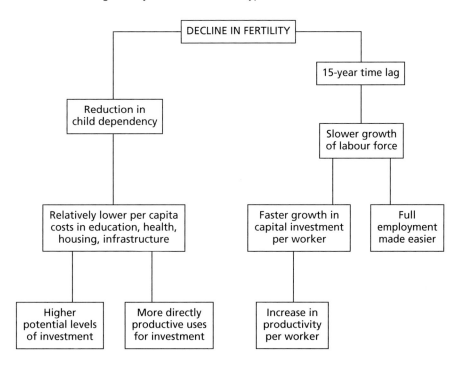

C Account for the time lag factor shown on the right-hand chain of causation in Figure 86. (*5L, 4M*)

D Discuss the range of economic advantages generally associated with a fall in fertility. (*10L, 6M*)

E Suggest reasons why these chains of causation might not materialise in reality in some countries. (*10L, 6M*)

68. The changing population structure of Brazil 1980–2010

Figure 87 shows population pyramids for Brazil.

A Describe how the population aged under 15 changes from one population pyramid to another. (*5L, 3M*)

B Explain this significant demographic change. (*6L, 5M*)

C Suggest how such a change in youth dependency is affecting Brazil. (*7L, 5M*)

D To what extent, and why, did the economically active population change in Brazil between 1980 and 2000? (*7L, 5M*)

E Describe and explain the changes in aged dependency between the four population pyramids. (*10L, 7M*)

Figure 87 Brazil: population pyramids 1980–2010

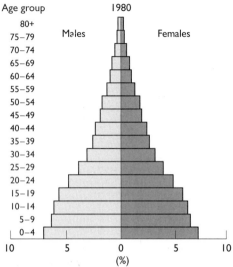

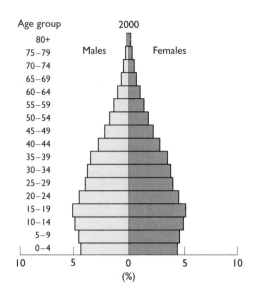

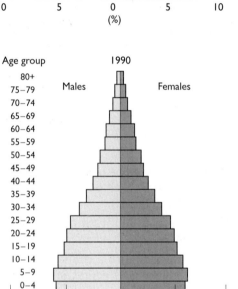

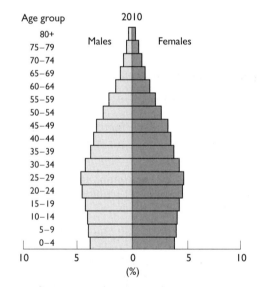

69. Population change and age structure in Inner London

Figure 88 shows changes in the population of Inner London.

A Describe the change in the population of Inner London from 1801 onwards. (*5L, 3M*)

B Suggest reasons for the changes you have described in A. (*7L, 5M*)

C Describe how the age structure of inner London's population changed between 1961 and 1991. (*6L, 5M*)

D What are the likely reasons for such changes? (*7L, 5M*)

E Discuss the likely economic and social impact of such changes in population structure on Inner London. (*10L, 7M*)

Figure 88

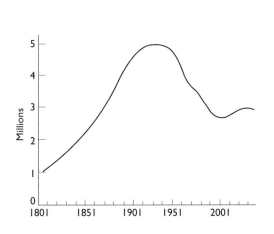

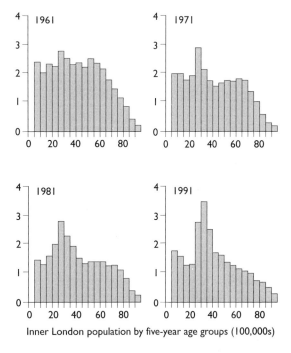

Inner London population by five-year age groups (100,000s)

70. The population structure of the United Kingdom

Figure 89 shows a population pyramid for the UK in 1996.

Figure 89

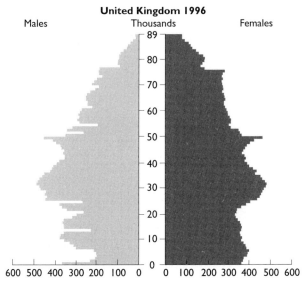

United Kingdom 1996

Source: Office for National Statistics; General Register Office for Scotland; Northern Ireland Statistics and Research Agency.

A Compare the number of males and females at the following ages:
   (i) under 1
   (ii) 30
   (iii) 50
   (iv) 80. (*6L, 4M*)

B Suggest reasons for the changes in sex ratio with age. (*5L, 4M*)

C What is the possible evidence of the impact of the First World War (1914–18) and the Second World War (1939–45) on the population structure of the UK? (*7L, 5M*)

D What is the evidence that the birth rate fluctuated considerably between 1966 and 1996? (*7L, 5M*)

E Compare the two components of the dependent population: child dependants (under 15) and elderly dependants (65 and over). How do you think the ratio between the two has changed over time? (*10L, 7M*)

Figure 90 Petersen's 'general typology of migration'.

Relation	Migratory force	Class of migration	Type of migration	
			Conservative	Innovating
Nature and man	Ecological push	Primitive	Wandering Ranging	Flight from the land
State (or equivalent) and man	Migration policy	Forced Impelled	Displacement Flight	Slave trade Coolie trade
Man and his norms	Higher aspirations	Free	Group	Pioneer
Collective behaviour	Social momentum	Mass	Settlement	Urbanisation

71. A typology of migration

A Look at Figure 90. What is the difference between migrations which are considered 'conservative' and those that are viewed as 'innovating'? (*5L, 3M*)

B Discuss two factors that might prompt primitive migration. (*5L, 4M*)

C Distinguish between forced migration and impelled migration. (*5L, 4M*)

D Under what circumstances might free migration develop into mass migration? (*8L, 6M*)

E What are the likely effects of a mass migration on the population structure of

(i) the donor country and

(ii) the receiving country? (*12L, 8M*)

72. The costs and returns from rural to urban migration

A With reference to Figure 91, explain two reasons for rural to urban migration. (*5L, 4M*)

B To what extent and why is rural to urban migration selective? (*6L, 4M*)

C Discuss the 'support costs' flowing from village to city. (*6L, 4M*)

D What are 'remittances'? Suggest how remittances are used in rural areas. (*7L, 5M*)

E Comment on the possible positive and negative effects of the other flows from city to village. (*11L, 8M*)

Figure 91 Simple model of the costs and returns from migration

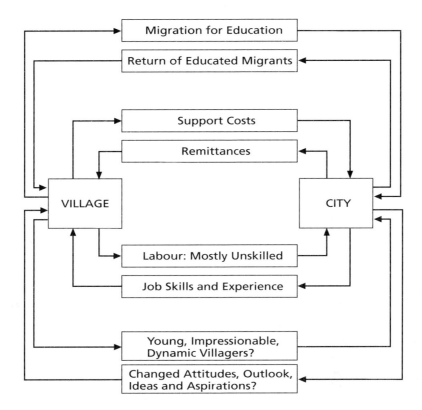

73. Mabogunje's systems approach to rural to urban migration

Figure 92 is a systems diagram for migration.

A Suggest how the 'rural control sub-system' might encourage or restrain potential migrants. (*6L, 4M*)

B Identify factors in the 'urban control sub-system' that might encourage a migrant to remain in the urban area or to return home. (*6L, 4M*)

C How would you expect the 'urban adjustment mechanism' to operate? (*7L, 5M*)

D Discuss the importance of feedback from urban areas to future out-migration from rural districts. (*7L, 5M*)

E 'The system and the environment in which it operates (the factors listed outside the rectangle) act and react upon each other continuously'. Discuss briefly. (*9L, 7M*)

Figure 92 A systems approach to migration [A. Mabogunje]

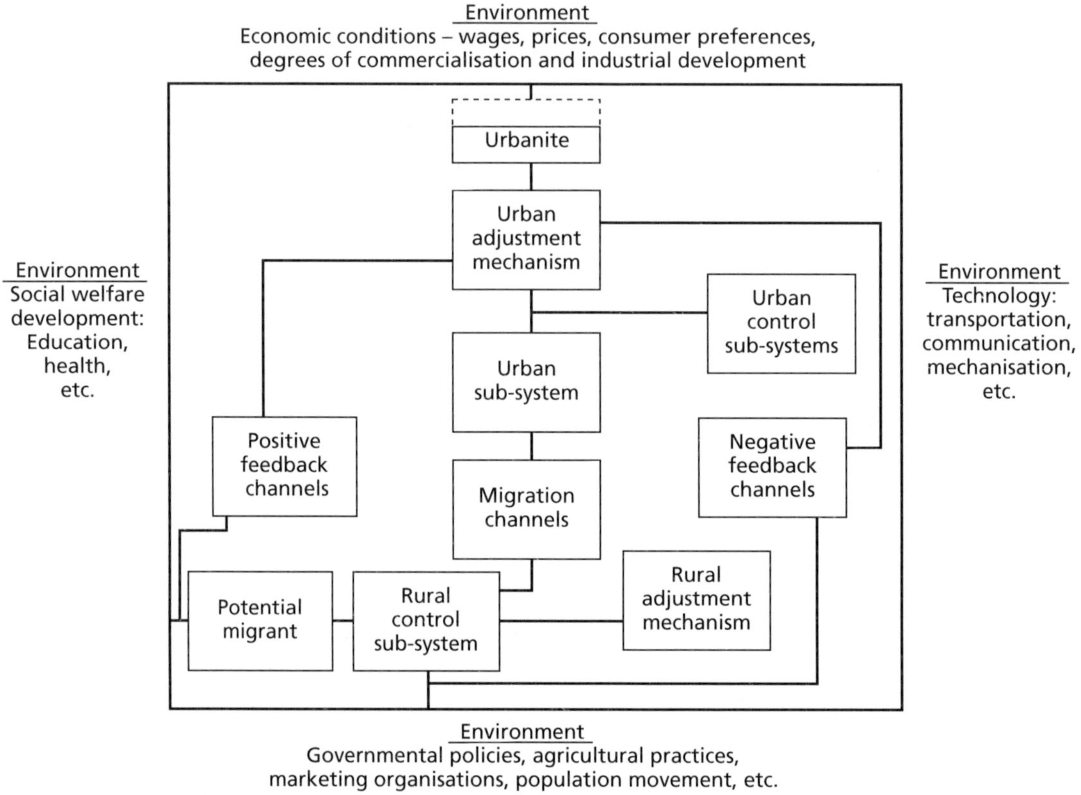

74. Policy approaches to internal migration

Figure 93 A typology of migration policy approaches

Policy approaches	Rationale
Negative	Emphasises the undesirability of migration and seeks to erect barriers to population movement and to forcibly 'deport' migrants
Accommodative	Accepts migration as inevitable, and seeks to minimise the negative effects in both origin and destination places
Manipulative	Accepts migration as inevitable and even desirable in some cases but seeks to redirect migration flows towards alternative destinations
Preventive	Rather than dealing with the symptoms of migration, attempts to confront the root causes by tackling poverty, inequality and unemployment at source, and reducing the attractiveness of urban areas to potential migrants

A Why have a number of countries attempted to reduce the rate of internal migration in the post-1945 period? (*6L, 4M*)

B Suggest two ways in which a negative policy approach could be implemented. (*7L, 4M*)

C Discuss ways in which an accommodative approach could reduce the negative effects of migration. (*7L, 5M*)

D Explain one planning measure that could be used to redirect migration flows towards alternative destinations (the manipulative approach). (*6L, 5M*)

E Outline the merits and limitations of the preventive approach to migration. (*9L, 7M*)

75. Population movement in the Third World

Figure 94 Spatial dimensions of population movement in the Third World [M. Parnwell]

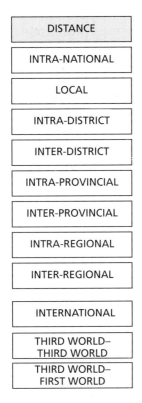

DISTANCE
INTRA-NATIONAL
LOCAL
INTRA-DISTRICT
INTER-DISTRICT
INTRA-PROVINCIAL
INTER-PROVINCIAL
INTRA-REGIONAL
INTER-REGIONAL
INTERNATIONAL
THIRD WORLD– THIRD WORLD
THIRD WORLD– FIRST WORLD

DIRECTION
RURAL–RURAL
RURAL–URBAN
URBAN–RURAL
URBAN–URBAN
PERIPHERY–CORE
CORE–PERIPHERY
TRADITIONAL– MODERN SPHERES

PATTERNS
STEP-MIGRATION
MIGRATION STREAM
COUNTER-STREAM

A What do you understand by the terms

(i) counter-stream

(ii) step-migration? (2 × [*3L, 2M*])

B Explain two possible reasons for rural to rural migration. (*5L, 4M*)

C Why would you expect migration from periphery to core to be greater than movement from core to periphery? (*6L, 4M*)

D Examine one aspect of migration that is not included in Figure 94. (*6L, 5M*)

E For one developing country you have studied, examine its human migration in terms of distance, direction and patterns. (*12L, 8M*)

76. Optimum population and improvements in technology

Figure 95

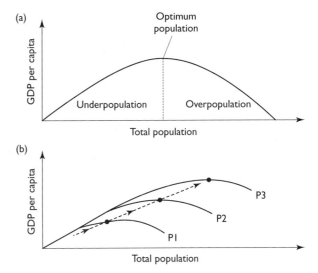

A Define the terms 'optimum population' and 'overpopulation'. ($2 \times$ [3L, 2M])

B Discuss the likely human characteristics of a country which is underpopulated. (6L, 4M)

C Explain the effect of improvements in technology on the optimum population. (6L, 4M)

D With reference to a country that considers its population/population growth to be too high, discuss what it is doing in terms of population policy. (8L, 6M)

E In Figure 95 the optimum population is thought of in terms of GDP per capita. Discuss other criteria that could be used to consider the concept of the optimum population. (9L, 7M)

77. World population and food supply

Figure 96 shows world population and food supply. Figure 97 shows the views of Boserup and Malthus.

A Describe the relationship between world population and food supply between 1951 and 1995. (5L, 3M)

B Explain three reasons for the increase in food production. (7L, 5M)

C Considering the substantial increase in global food production, why do so many people in developing countries not have enough to eat? (7L, 5M)

D Which view of the relationship between population growth and food supply in Figure 97 is supported by Figure 96? Justify your answer. (8L, 6M)

E Suggest why some experts are pessimistic about the future relationship between population growth and food supply. (8L, 6M)

Figure 96

Figure 97 Boserup's and Malthus's views of population growth and food supply

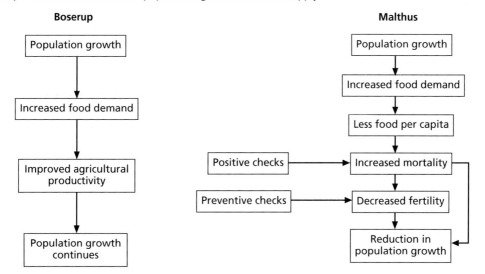

78. World population and water supply

Figure 98 is a newspaper report, 8 November 2001.

A Why does the United Nations expect the global water supply problem to become much worse over the next 50 years? (*6L, 4M*)

B Why will the greatest problems be experienced by LEDCs? (*7L, 5M*)

C Explain the consequences in LEDCs of

(i) poor water supply in terms of quantity and

(ii) poor water supply in terms of quality. (2 × [*5L, 4M*])

D For an MEDC you have studied in relation to water problems:

(i) explain the causes of these problems and

(ii) discuss the possible solutions to them. 2 × [*6L, 4M*])

Figure 98

World 'to run out of water in 50 years'

BY CHARLES CLOVER

ENVIRONMENT EDITOR

THE WORLD will begin to run out of fresh water by 2050 because of the expected population growth to 9·3 billion, the UN Population Fund said yesterday.

All of the projected growth, from the present population of 6·1 billion, will be in developing countries already straining to feed and provide basic services to their people.

Populations of the 49 least developed countries will triple in size from 668 million to 1·86 billion. Water supplies are already stretched in the poorest countries and water use has grown six-fold in 70 years.

Worldwide, some 54 per cent of the annual available fresh water is being used, two-thirds for agriculture. By 2025, it could be 70 per cent because of population growth. If consumption everywhere reached the level of developed countries, it would be 90 per cent.

The fund's report, State of the World's Population 2001, said the Earth's resources were being used at a greater intensity than at any time in history. Since 1960, world population had doubled.

Last year 508 million people lived in 31 water-stressed countries but, by 2025, three billion will be in 48 such countries. By 2050, 4·2 billion will be in countries unable to meet the UN daily requirement of 50 litres of water a person for drinking, washing and cooking.

Source: Daily Telegraph 8/11/01

79. Distinguishing between urban and rural settlement

Figure 99 shows the hierarchy of settlement.

Figure 99

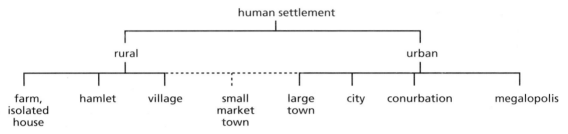

Small market towns, which have extremely strong links to the surrounding countryside, are considered rural according to some classifications but urban according to others.

A What do you understand by the terms 'conurbation' and 'megalopolis'? (2 × [*3L, 2M*])

B Briefly discuss three criteria that can be used to distinguish between urban and rural settlement. (3 × [*3L, 2M*])

C Why are small market towns the most difficult to classify in the rural–urban divide? (*4L, 3M*)

D Describe and explain the relationship between settlement size and the number of settlements. (*6L, 5M*)

E Why might the hierarchy of settlement in a region change over time? (*10L, 7M*)

80. Rural depopulation

Figure 100 Model of the downward spiral of rural depopulation

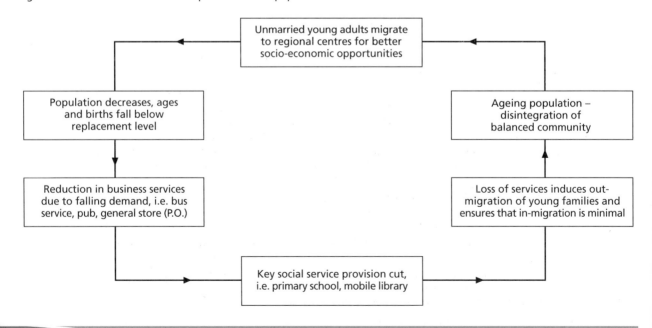

Figure 101 Depopulated village in northern Spain

A What is the visual evidence in the photograph (Figure 101) that this village in northern Spain has undergone depopulation? (*4L, 3M*)

B Why is out-migration from rural areas dominated by young adults? (*6L, 4M*)

C Suggest reasons why privately owned services usually close in rural areas before publicly owned services. (*7L, 5M*)

D Examine the impact of cuts in service provision. (*8L, 6M*)

E How might local or regional government try to reverse such a trend? (*10L, 7M*)

81. The rural transport problem

A What was the initial reason for the increase in car ownership in rural areas? (*4L, 3M*)

B Explain two direct consequences of this change. (*5L, 4M*)

C Why have many public transport services in rural areas been cut or reduced in frequency? (*8L, 5M*)

D Which sectors of the rural population have been most affected by the reduction in public transport? (*8L, 5M*)

E Discuss one strategy that local government could adopt to try to prevent further reductions in rural public transport. (*10L, 8M*)

Figure 102 The decline of public transport in rural areas

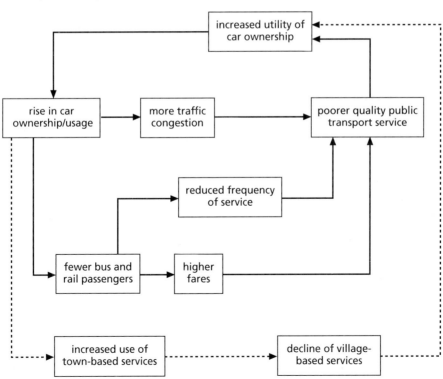

Source: M. J. Moseley *Accessibility: the rural challenge*, Methuen and Co. 1979

82. Key settlements in rural areas

Figure 103 shows the concept of key settlements.

Figure 103

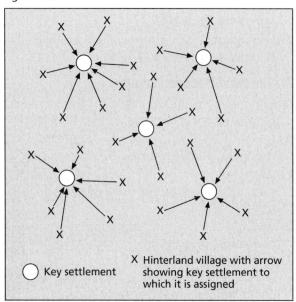

○ Key settlement

X Hinterland village with arrow showing key settlement to which it is assigned

A What is a 'key settlement'? (*3L, 2M*)

B Why have a number of counties in Britain adopted a key settlement policy at various times over the past 40 years or so? (*6L, 4M*)

C In a relatively isolated rural area, which villages would most likely be designated key settlements? (*6L, 5M*)

D Suggest why some rural communities in an area might be against the introduction of a key settlement policy. (*8L, 6M*)

E In an area where a key settlement policy has been running for some time, what would the likely demographic, social and economic differences be between villages which were key settlements and villages which were not? (*12L, 8M*)

83. Settlement patterns in Britain

A Define the terms 'dispersed settlement pattern' and 'nucleated settlement pattern'. (2 × [*3L, 2M*])

B Describe the distribution of rural settlement patterns in Britain (Figure 104). (*6L, 4M*)

C Under what circumstances does a nucleated settlement pattern tend to develop? (*6L, 5M*)

D Discuss the main reasons for the evolution of a dispersed pattern of settlement. (*6L, 5M*)

E Identify the problems encountered by the rural population in recent decades where a dispersed settlement pattern exists. (*11L, 7M*)

Figure 104 Rural settlement in Britain

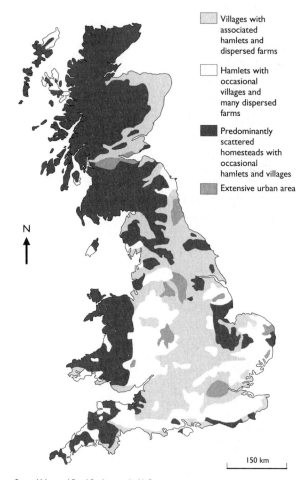

☐ Villages with associated hamlets and dispersed farms

☐ Hamlets with occasional villages and many dispersed farms

■ Predominantly scattered homesteads with occasional hamlets and villages

▨ Extensive urban area

N
↑

150 km

Source: Urban and Rural Settlements by H. Carter

84. Service provision in rural areas

A Using the evidence provided in the four photographs (Figures 105 to 108), assess the extent of service provision in this mid-Surrey village. (*6L, 4M*)

B What don't the photographs tell you about the services you have identified? (*6L, 5M*)

C Discuss the factors that are likely to have affected the level of service provision in this village. (*8L, 6M*)

D Discuss the primary and secondary data you would need to acquire to assess the ways in which the village had changed over the past 30 years. (*15L, 10M*)

Figure 105 Leigh, Surrey

Figure 106 Leigh, Surrey

Figure 107 Leigh, Surrey

Figure 108 Leigh, Surrey

85. Metropolitan villages

Figure 109 shows the morphology of metropolitan villages.

A Suggest a significant reason for the site of the original core village. (*4L, 3M*)

B Why were isolated dwellings often located within easy reach of such villages? (*5L, 4M*)

C Explain the changes shown in Stage 2. (*6L, 4M*)

D Account for the more complex development of the village shown in Stage 3. (*8L, 6M*)

E Compare the morphology of a village you have studied with Figure 109. (*12L, 8M*)

Figure 109

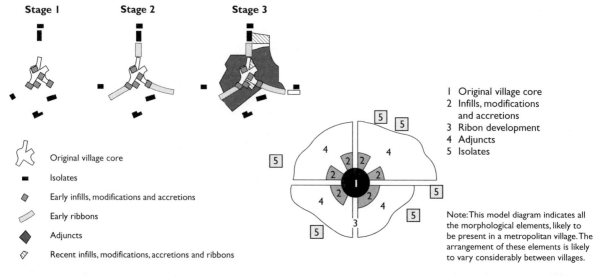

Stage 1   **Stage 2**   **Stage 3**

Original village core

Isolates

Early infills, modifications and accretions

Early ribbons

Adjuncts

Recent infills, modifications, accretions and ribbons

1 Original village core
2 Infills, modifications and accretions
3 Ribon development
4 Adjuncts
5 Isolates

Note: This model diagram indicates all the morphological elements, likely to be present in a metropolitan village. The arrangement of these elements is likely to vary considerably between villages.

(a) Possible stages of morphological evolution of a suburbanised village

(b) Metropolitan village: morphological features

86. The second homes issue

A Why has the number of 'second homes' increased in Britain in recent decades? (*5L, 3M*)

B Which type of rural areas have experienced the greatest increase in second home ownership? (*5L, 4M*)

C What are the disadvantages of the growth in the number of second homes to the established population? (*8L, 5M*)

D Suggest two possible benefits to the local economy of the growth in the number of second homes. (*6L, 5M*)

E Discuss the measures the Exmoor National Park Authority would like to introduce to curtail the increase in second home ownership. (*11L, 8M*)

Figure 110

In a bid to keep Exmoor cottages out of the hands of outsiders the National Park Authority wants rules to prevent new homes being bought for holiday and weekend use.

THE parish of Exton and Bridgetown is tucked away in a steeply wooded fold of Exmoor where one in three of the cottages is a weekend home. Fifteen miles away on the seaward side of the stunning moor, in the parish of Brendon virtually every other house is a holiday cottage and young local people find themselves living at home with their parents into their thirties or simply leaving.

Houses prices in Exmoor rocketed by 26 per cent last year as town-weary newcomers sought the perfect cottage hideaway. The price of two-bedroom homes has risen by around £30,000 to £115,400 according to the latest figures, putting them completely out of reach of all but a few Exmoor newlyweds. It's against this background that the Exmoor National Park Authority feels compelled to act as the moorland folk, already battered by BSE, foot-and-mouth and the threat to hunting, are being "weekender cleansed". National Park planning officer Jack Ellerby says: "A recent survey put the second homes problem and the fact that local youngsters cannot even dream of buying a house in the area as the concern that troubles most.

"Statistically Brendon and Exton parishes have the highest proportion of second home blighting and it's interesting that when the numbers started to rise rapidly in the early Nineties the local shops in both villages were killed off. "We believe that's down to the Sainsbury's syndrome where weekenders come in with bags full of groceries and supplies they've bought elsewhere. If enough of the houses in a village are doing this then the village shop dies.

"We have no doubt this lifestyle has a massively negative impact on local services, on community life and on the ability of local people to make a life for themselves. We now have young people living in Minehead or Tiverton on the edge of the moor and commuting back to work on farms. Others are living with their parents well into their thirties.

"It's not an Exmoor problem alone. Most of the other National Park areas have the same problem and in some places like Wales, the Lakes and the Yorkshire Dales you have villages with nearly 70 per cent of the housing being holiday homes." Mr Ellerby wants to stem the flow by insisting that holiday homes should be seen as a change of use from a permanent dwelling and therefore should need planning consent.

He would also like to see the size of new-build houses in the National Park restricted to 90 square metres hoping that would downsize the market value as well. Any attempt to ban outsiders from buying homes could face an intervention from Stephen Byers, Secretary of State for Transport, Local Government and the Regions. He warned such a policy would be open to legal challenge from prospective homeowners.

Source: Evening Standard 6/9/2001.

87. Rural poverty in LEDCs

A According to the United Nations, how many people in the world live in extreme poverty? (*1L, 2M*)

B Suggest why so many of the world's poor live in rural areas. (*6L, 4M*)

C Comment on the geographical distribution of rural poverty. (*6L, 5M*)

D Discuss the solutions to rural poverty proposed by IFAD. (*10L, 6M*)

E Why is IFAD particularly keen to improve the lives of rural women? (*12L, 8M*)

Figure 111

Helping the dirt-poor

ROME

BOLD aims are easy enough: getting things done is the hard bit. The United Nations' millennium summit last year resolved to halve the number of the world's poor by 2015. Five years earlier, UN members had agreed to work out programmes to end the worst sorts of poverty. Yet the rate of poverty reduction in the past decade was less than a third of what is needed to achieve the UN's goal, says a report published on February 6th by the International Fund for Agricultural Development (IFAD), a Rome-based agency that tries to help the poorest of the poor.

About 1.2 billion people live in extreme poverty, meaning on less than a dollar a day. Three-quarters of them live and work in rural areas. So, says IFAD, it is necessary to concentrate on the countryside, and on what can be done to make poor farmers more efficient.

Sticking its neck out, IFAD wants to see an expansion of the use of genetic technology in farming, which it reckons will increase crop yields and reduce disease among animals and plants. It also argues that a redistribution of land into more equal holdings, especially small family farms, will help efficiency. Less controversially, it calls for an improvement in the distribution of water to the rural poor, and for better roads so that they can get their products to the market more easily. And, of course, it says it is essential to have more education and better health-care.

The agency is particularly keen on improving the lot of rural women. They are generally poorer than the men, it says, less educated and in worse health, own no land and die sooner. "There is an enormous case for investing in women," says Eve Crowley, who advises IFAD on the subject. To put it crudely, women are an under-used resource. And the fact that in many places they have no access to credit hampers rural development.

Rural blight

Regional distribution of dollar-a-day rural poverty

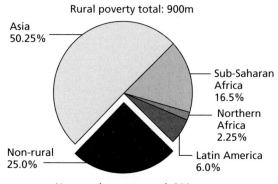

Rural poverty total: 900m

Asia 50.25%

Sub-Saharan Africa 16.5%

Northern Africa 2.25%

Latin America 6.0%

Non-rural 25.0%

Non-rural poverty total: 300m

Source: The Economist 10/2/01

88. The overlap of rural and urban deprivation

Figure 112 shows the overlap of rural and urban deprivation.

A What do you understand by the term 'deprivation'? (*4L, 2M*)

B Explain two problems that Figure 112 applies to peripheral rural areas only? (*7L, 5M*)

C Examine two problems that are common to both inner urban and peripheral rural areas. (*7L, 5M*)

D Discuss two problems which Figure 112 applies to inner urban areas only. (*7L, 5M*)

E Suggest why deprived urban areas have received a much higher level of government funding in recent decades than deprived rural areas. (*10L, 8M*)

Figure 112

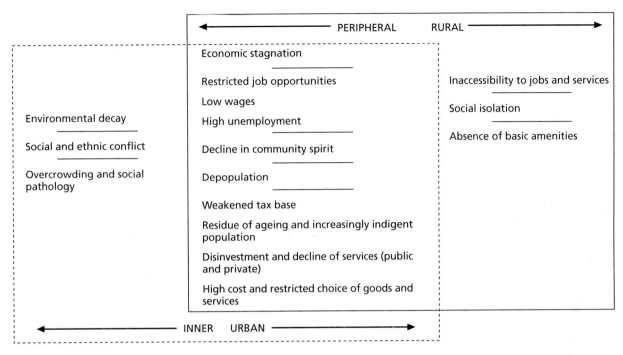

89. Global variations in urbanisation

Figure 113 shows the percentage of the world's urban population.

Figure 113 Percentage of population that was urban in 1995

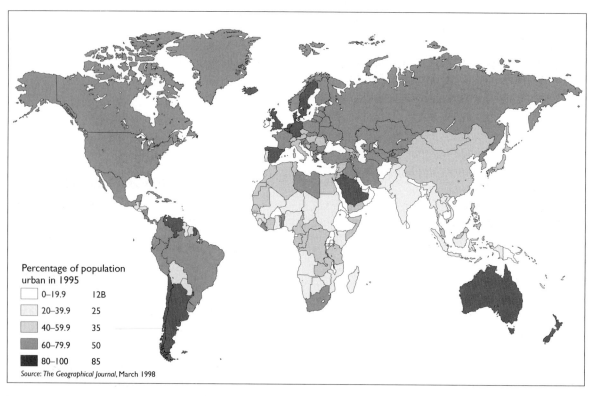

Percentage of population urban in 1995

	0–19.9	12B
	20–39.9	25
	40–59.9	35
	60–79.9	50
	80–100	85

Source: The Geographical Journal, March 1998

A Describe the variations in the level of urbanisation by continent. (*5L, 3M*)

B Why is the level of urbanisation higher in the developed world than in the developing world? (*6L, 4M*)

C For the developing world suggest reasons why urbanisation is more advanced in Latin America than in either Asia and Africa. (*6L, 5M*)

D Explain why Asia has a relatively low level of urbanisation overall yet more urban centres over 5 million population (Figure 114) than any other continent. (*8L, 6M*)

E Account for the global trends in the distribution of very large urban areas shown in Figure 114. (*10L, 7M*)

Figure 114 Urban centres with populations of five million or more

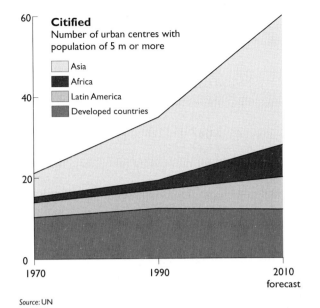

Source: UN

Figure 115 The urbanisation curve

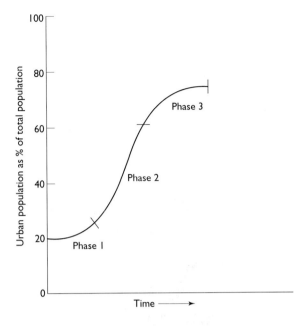

90. The cycle of urbanisation

A Describe the differences between the three stages illustrated in Figure 115. (*6L, 3M*)

B On a 1:50 000 Ordnance Survey map of an urban area in Britain, what evidence might you expect to find for

(i) stage 1

(ii) stage 2? (2 × [*3L, 2M*])

C At what stage are most LEDCs? Give reasons for your answer. (*5L, 4M*)

D For what reasons do countries move from Stage 2 to Stage 3? (*8L, 6M*)

E On Figure 115 add a fourth stage to show counterurbanisation. What is the evidence that some countries are now in this fourth stage? (*10L, 8M*)

91. Counterurbanisation and reurbanisation

A Discuss the evidence for counterurbanisation provided by Figure 116. (*6L, 4M*)

B Examine the reasons for this process in MEDCs. (*6L, 4M*)

C What impact has counterurbanisation had on

(i) metropolitan areas and

(ii) smaller urban and rural areas? (2 × [*5L, 4M*])

D What do you understand by the term 'reurbanisation'? (*3L, 2M*)

E Discuss the reasons for the occurrence of this process in some MEDC cities in the 1990s. (*10L, 7M*)

Figure 116 Net within-Britain migration, 1990–91, by district types

District type	Population 1991	Net migration 1990–91	%
Inner London	2 504 451	−31 009	−1.24
Outer London	4 175 248	−21 159	−0.51
Principal metropolitan cities	3 992 670	−26 311	−0.67
Other metropolitan districts	8 427 861	−6 900	−0.08
Large non-metropolitan cities	3 493 284	−14 040	−0.40
Small non-metropolitan cities	1 861 351	−7 812	−0.42
Industrial districts	7 475 515	7 194	0.10
Districts with new towns	2 838 258	2 627	0.09
Resorts, ports and retirement districts	3 591 972	17 637	0.49
Urban–rural mixed	7 918 701	19 537	0.25
Remote urban–rural	2 302 925	13 665	0.59
Remote rural	1 645 330	10 022	0.61
Most remote rural	4 731 278	36 450	0.77

Note:
Metropolitan cities and districts includes the Central Clydeside conurbation area
Source: Calculated from the 1991 Census SMS and LBS/SAA (ESRC/JISC purchase) Reproduced from *Population Trends 83*, Spring 1996

92. Mann's model of a British city

Figure 117 Mann's model of a British city

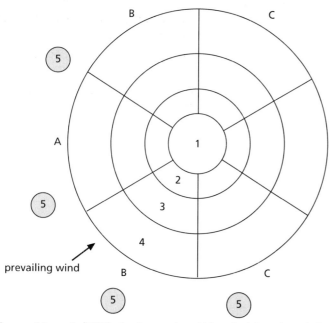

1 Central business district
2 Transitional zone
3 Zone of small terrace houses in sectors C, D
   Larger by-law housing in sectors B
   Large old houses in sector A
4 Post-1918 residential areas with post-1945 development mainly on the periphery
5 Commuting distance 'dormitory' towns

A Middle-class sector
B Lower middle-class sectors
C Working-class sectors (and main council estates)
D Industry and lowest working-class sector

prevailing wind

Source: Mann, P. (1965), *An Approach to Urban Sociology* (Routledge & Kegan Paul)

A On which two previous theories of urban land use is Mann's model based? (*2L, 2M*)

B Explain the central location of the CBD. (*4L, 3M*)

C Suggest two reasons for the location of the industry and lowest working-class sector. (*5L, 3M*)

D Outline the characteristics of, and processes operating in, the transitional zone. (*8L, 5M*)

E Which factors might explain the location of the middle-class sector? (*8L, 6M*)

F Discuss the likely characteristics of the commuting distance dormitory towns. (*8L, 6M*)

93. A new model of the Latin American city

Model 1 (Figure 118) was produced in 1980. Model 2 (Figure 119), by one of the original authors, was produced as an updated version of the original in 1996.

Figure 118 Model 1: The Latin American City 1980

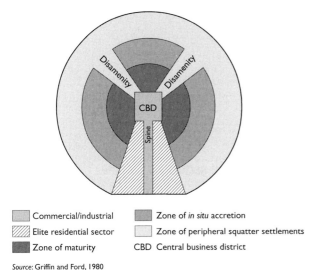

▢ Commercial/industrial	▨ Zone of *in situ* accretion
▨ Elite residential sector	▢ Zone of peripheral squatter settlements
▨ Zone of maturity	CBD Central business district

Source: Griffin and Ford, 1980

Figure 119 Model 2: The Latin American City, mid-1990s

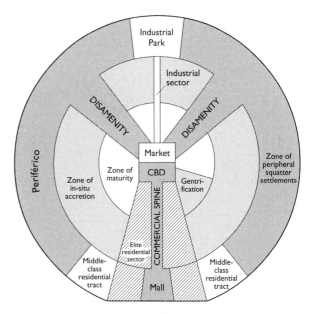

A Discuss the characteristics that are similar to both models. (*8L, 6M*)

B Explain the location of the Industrial Park in Model 2. (*6L, 4M*)

C What do you understand by the term 'gentrification'? Comment on the location of this zone in Model 2. (*6L, 5M*)

D Comment on the likely characteristics of the Mall. (*6L, 4M*)

E Suggest reasons for the development of a middle-class residential tract on the edge of the city. (*9L, 6M*)

94. A model of the North American rural–urban fringe

Figure 120 The rural–urban fringe

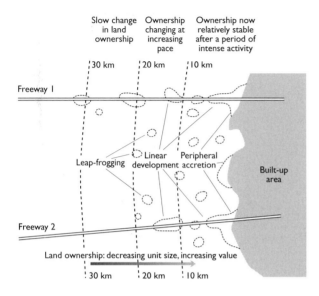

Source: *North America: A Human Geography*, by P. Guinness and M. Bradshaw, Hodder & Stoughton.

A Briefly explain the following in the context of the model:

  (i) peripheral accretion,

  (ii) linear development,

  (iii) leap-frogging. (3 × [*3L, 2M*])

B Suggest why units of land ownership decrease in size but increase in value with proximity to the built-up area. (*6L, 4M*)

C Account for the contrasting rates of change in the three zones shown in the model. (*6L, 4M*)

D What effects might imminent urbanisation have on agriculture in the rural–urban fringe? (*4L, 3M*)

E Discuss the planning measures that could be introduced to prevent the chaotic sprawl of the urban area. (*10L, 8M*)

95. Settlement size and range of functions

A Describe the relationship between settlement population size and the number of functions provided. (*4L, 3M*)

B Explain the reasons for such a relationship. (*6L, 4M*)

C Suggest reasons for the two anomalies shown in Figure 121. (*6L, 4M*)

D How might a settlement change its position in the hierarchy shown in Figure 121? (*7L, 4M*)

E For a settlement you have studied describe and explain the level of its service provision. (*12L, 10M*)

Figure 121 Relationship between settlement size and range of functions

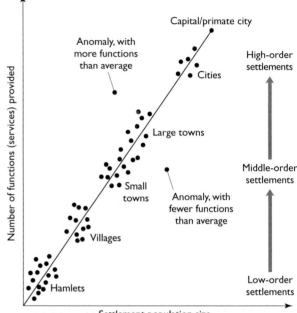

96. Hierarchy of settlement patterns

Figure 122 shows the hierarchy of settlement patterns.

A What is meant by the 'rank-size rule'? (*4L, 3M*)

B Why is a log. scale generally used to illustrate settlement hierarchies? (*4L, 4M*)

C How does the stepped order pattern differ from the rank-size rule? (*5L, 4M*)

D With reference to an example, explain why a primary pattern of settlement sometimes occurs. (*11L, 7M*)

E With reference to an example, explain why a binary pattern of settlement may occur. (*11L, 7M*)

Figure 122

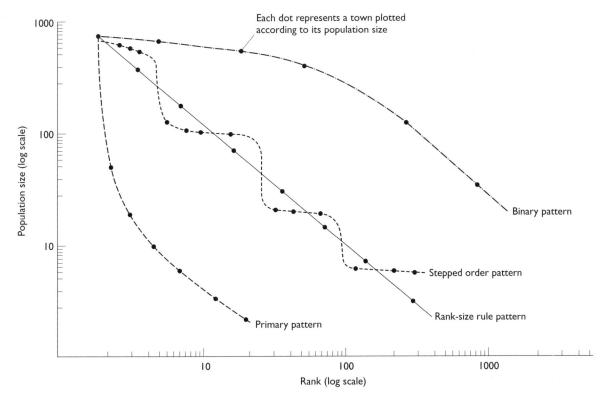

Each dot represents a town plotted according to its population size

Population size (log scale)

Binary pattern

Stepped order pattern

Rank-size rule pattern

Primary pattern

Rank (log scale)

97. The gravity model

Figure 123 Reilly's gravity model

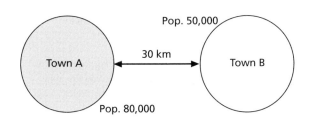

Pop. 50,000

30 km

Town A ⟷ Town B

Pop. 80,000

Breaking point AB =

$$\text{Breaking point AB} = \frac{\text{Distance from A to B}}{1 + \sqrt{\dfrac{\text{Population A}}{\text{Population B}}}}$$

A Using the formula for Reilly's gravity model given in Figure 123, calculate the breaking point between the two towns. (*4L, 3M*)

B Outline the theoretical basis of the theory. (*6L, 4M*)

C Discuss the factors which in reality could result in a breaking point different from your answer to A. (*6L, 5M*)

D Explain three ways in which you could attempt to determine the breaking point between two settlements through fieldwork. (*9L, 6M*)

E Why might major retail outlets be interested in this concept? (*10L, 7M*)

98. Urban building height profile

Figure 124 is a photograph of Seattle, USA.

A Describe the distribution of buildings by height in Figure 124. (*5L, 4M*)

B Which land uses would you expect to occupy the cluster of very tall buildings, and why? (*6L, 4M*)

C Which land uses are likely to occupy zones of lower building height, and why? (*6L, 4M*)

D What are the main reasons for such considerable variations in building height? (*8L, 6M*)

E Discuss the factors that might cause the height profile of Seattle to change in the future. (*10L, 7M*)

Figure 124

99. Developing greenfield and brownfield sites

Figure 125 Proposed location of new housing in England and Wales, 1996–2016

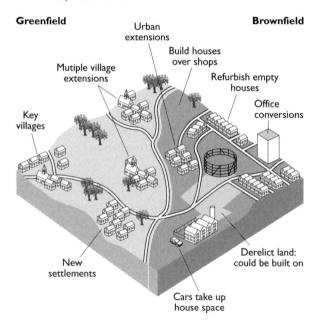

A Define the terms 'greenfield' and 'brownfield'. (2 × [*2L, 2M*])

B Why do property developers prefer generally to build on greenfield sites? (*6L, 3M*)

C Discuss the merits and limitations of the different ways in which new housing could be provide within urban areas. (*8L, 5M*)

D What are the main arguments against greenfield development in general? (*7L, 6M*)

E Examine the advantages and disadvantages of the different types of greenfield development identified in the diagram. (*10L, 7M*)

100. Central urban congestion charges

Figure 126A

Car zone 'will hit house prices'

By Ben Webster

Transport Correspondent

PROPERTY prices inside Ken Livingstone's London congestion charge zone will rise by up to £40,000 because residents will receive a 90 per cent discount on the fee, estate agents say.

Homeowners just outside the zone, however, will see the value of their properties fall because of rat-running motorists, rogue parking and the need to buy a £1,250 annual pass to enter the zone.

The London Mayor, who is introducing the £5 daily charge in January 2003 for vehicles entering the zone between 7am and 7pm, Monday to Friday, hopes to cut congestion in the capital by 10 to 15 per cent.

Dozens of towns and cities across the United Kingdom are waiting to judge the success of London's scheme before deciding whether to introduce their own form of congestion charging.

On the roads that make up the zone's border, people living on one side of the street will get the full discount, while those on the opposite side will have to pay the full charge.

George Franks, the manager of Douglas and Gordon estate agents, which has its offices just outside the zone in Battersea, said that people considering making an offer for a home in Central London would be wise to check where it is in relation to the congestion charge border. "If you are on the wrong side of it, it is going to hurt," he said.

"The charge will bring a huge correction in the market. For people inside the zone it will be like having the freedom of the City of London: being able to drive wherever you like."

Residents across the border unwilling to pay the charge will face £80 fines or severe restrictions on where they can drive and park. "You will only be able to go out of your door and turn left, not right; otherwise you will have to pay £1,250," Mr Franks said.

Chris Lee, of Felicity Lord agents at Tower Bridge, likened the impact of the charge to the arrival of the Jubilee Line in parts of South London, where property prices shot up in the late Nineties by 20 per cent.

He said that a £200,000 two-bedroom flat inside the zone would rise by 15 to 20 per cent because of the £1,125 saving on the cost of an annual pass.

"You can multiply that sum over several years, and then there's always the possibility of the charge going up," Mr Lee said.

Mr Livingstone has pledged not to increase the charge in his first term of office, but there has been speculation that it could rise to as much as £20 a day if he is re-elected in mid-2004.

Source: The Times 23/7/01

A What are congestion charges? (*4L, 2M*)

B Why have more and more urban areas considered introducing congestion charges? (*5L, 4M*)

C Using Figure 126 only, discuss the rationale of the congestion charge boundary. (*6L, 5M*)

D What advantages do you think congestion charges will bring to London in general

Figure 126B Map of proposed congestion charge area in Central London

CONGESTION CHARGE BOUNDARIES

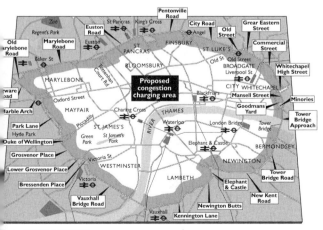

Source: The Times 23/7/2001.

(social benefits) and to certain people in particular (individual benefits)? (*10L, 7M*)

E What are the possible social costs and individual costs that might result from congestion charges? (*10L, 7M*)

101. London: the distribution of low-income households

A Describe the distribution of residents in receipt of income support in 1994. (*5L, 3M*)

B Explain the main reasons for this pattern of distribution. (*6L, 4M*)

C To what extent had the distribution changed from the earlier maps (1983 and 1987)? (*6L, 5M*)

D How far does Figure 128 help to explain the considerable increase in the number of people in receipt of Income Support? (*8L, 6M*)

E For what reasons might the distribution of low-income households in London change in the future? (*10L, 7M*)

Figure 127 Gross hourly (male) earnings (£). Greater London employees, 1979–95. Not adjusted for inflation

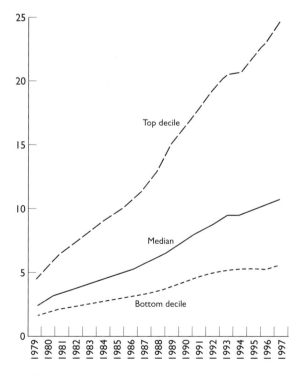

Source: New Earnings Survey.

Figure 128 Percentage of residents in receipt of SB/IS, London boroughs 1983–94

Figure 129 Inner-city decline – the downward spiral

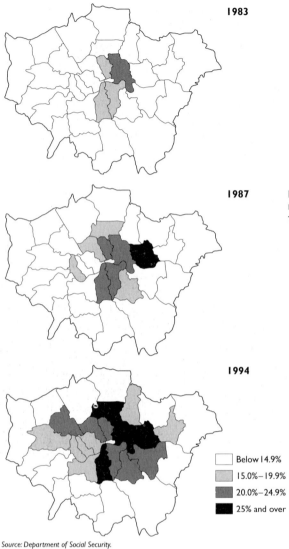

1983

1987

1994

Below 14.9%

15.0%–19.9%

20.0%–24.9%

25% and over

Source: Department of Social Security.

The proportion of Londoners reliant upon means-tested benefits has grown dramatically since the early 1980s. Just below 1 million residents were in receipt of Income Support (IS) in 1994 (17% of the adult population) compared with over 500,000 in 1989 (11%). Including the partners and children of those receiving benefit, over 1.5 million Londoners were reliant upon IS in 1994. While changes to the social security system in 1988 mean that data are not strictly comparable, as means-tested benefits they are strongly indicative of trends over the period.

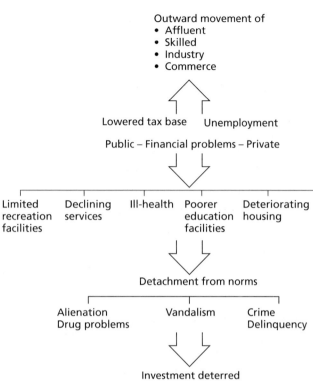

Source: A Modern Dictionary of Geography, M. Witherick et al, Arnold, 2001

B Discuss the financial problems encountered by both the public and private sectors as a result of inner-city decline. (*5L, 4M*)

C Examine the relationship between the symptoms of inner-city decline and 'detachment from norms'. (*8L, 5M*)

D To what extent does housing in inner cities generally differ from that in the suburbs? (*7L, 5M*)

E For an inner-city area you have studied, discuss the efforts undertaken to improve the quality of the urban environment. (*10L, 8M*)

103. Renovation of an inner-city estate

Figures 130 and 131 show the White City Estate, inner London. It was built in the 1930s and was renovated in the 1990s.

A Using the evidence provided in Figures 130 and 131 only, describe three ways in which the design of this block of flats has been

102. Inner-city decline: the downward spiral

A What were the initial reasons for decentralisation from the inner city in the early part of the twentieth century? (*5L, 3M*)

modified to reduce crime and vandalism. (*6L, 3M*)

B Explain why such measures have proved to be effective on many local authority estates. (*8L, 6M*)

C From the photograph identify one way in which each individual flat has been modernised. (*3L, 2M*)

D This is one of the most deprived local authority housing estates in West London. How is such deprivation measured by central and local government? (*8L, 6M*)

E Changes in urban design can improve the quality of life of residents to a certain extent. What needs to happen in the wider urban environment to generate more far-reaching improvements to the quality of life of residents on deprived estates? (*10L, 8M*)

104. Gentrification and filtering

A Define the terms 'gentrification' and 'filtering'. (2 × [*3L, 2M*])

B What is the evidence in Figure 133 that gentrification is taking place in this street in inner London? (*4L, 3M*)

C What type of inner-city area has been most susceptible to gentrification? (*7L, 5M*)

D What evidence would you look for in successive census reports to show that gentrification had taken place in an area? (*8L, 6M*)

Figure 130 Renovated front access

Figure 131 Redesigned rear space

Figure 132 The processes of filtering and gentrification

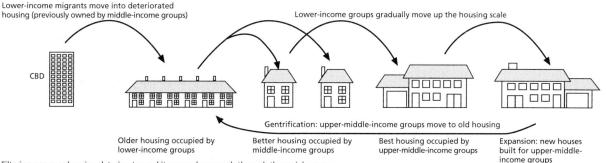

Source: Advanced Geography Revision Handbook, by G. Nagle & K. Spencer, OUP

E Discuss the advantages and disadvantages of gentrification to inner-city areas. (*10L, 7M*)

Figure 133 Gentrification in inner London

105. Urban managers

Figure 134 Urban managers and some of their responsibilities

A What is meant by the term 'urban manager'? (*3L, 2M*)

B Discuss the role that national government plays in the management of individual urban areas. (*6L, 4M*)

C How does urban government obtain its funding? What does it do with this money? (*8L, 6M*)

D Discuss the role of 'gatekeepers' in the urban management system. (*8L, 5M*)

E Examine the role of employers, service providers and developers in urban management. (*10L, 8M*)

106. Squatter settlements

Figure 135 The hazards in squatter developments

Living dangerously

DEVELOPING countries generally have bigger environmental problems than the rich world, but some places are a lot worse than others. The ungreenest spots of all are squatter settlements. Known as *favelas* in Brazil, *barrios* in Venezuela and *jhuggie* settlements in India, these slums are similar the world over. They are made of flimsy materials such as cardboard and scrap metal, built on hazardous land, poorly served with clean water and sewerage, and piled high with rubbish. They are also huge, and still growing fast. In Asia, according to one estimate, a quarter of the urban population lives in slums.

The way they come to be there is much the same the world over too. Typically, poor migrants arrive from the countryside. Unable to afford legal housing in the city, they squat on land where they will meet least resistance from landowners: steep hillsides, river beds, railway cuttings or sites next to industrial plants. Such places can be dangerous. When a hurricane hit Acapulco, in Mexico, last October, hundreds of shacks were swept away with their inhabitants inside them. In Rio de Janeiro and Caracas, thousands of homes perched on hillsides are washed away by floods and mudslides every year. Many victims of the accident in 1984 at a chemical plant in Bhopal, India, in which poison gas killed around 3,000 people, were living in nearby slums.

Once settled, squatter communities usually start lobbying the local government for legal recognition. This is often a long and tortuous process which becomes tangled in corruption. In parts of Mexico, for example, squatters make regular payments to local politicians in return for promises that they will not be evicted, and for moves towards official recognition. Yet until the settlements are on a proper legal footing, they are much less likely than other areas to benefit from municipal services such as electricity, water supply, sewerage and rubbish collection.

People in squatter communities also face a host of other ills, including high levels of crime, drug-taking and malnutrition. But unsanitary conditions are among the main reasons for high death rates. A study of Tondo, a squatter settlement in Manila, found that infant deaths were three times the level in legal settlements in the city.

How can the environmental problems of squatter communities be solved? Arguably municipal governments should be quicker to invest in basic services; but they are strapped for cash, and when they have it, they often waste it. A subtler solution is to speed up the process of legal recognition. If slum dwellers know they may be evicted at any time, they are unlikely to invest much in their homes. Conversely, slum projects in Latin America and Asia suggest that granting secure tenure prompts locals to invest in safer homes and better sanitation. That does not mean squatters should automatically be given ownership of land they occupy; but it does raise hopes that, once a government has settled the issue of land title, some of the worst environmental problems may disappear.

Source: The Economist 21/3/98

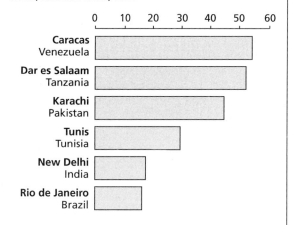

Squatter housing as a % of total housing stock, selected cities, 1990

Source: World Resources Institute

A Suggest reasons for the differing extent of squatter housing in the six cities shown in Figure 135. (*5L, 3M*)

B Why are health conditions much worse in squatter settlements than elsewhere in urban areas? (*6L, 4M*)

C Why is legal recognition so important to the improvement in living conditions in squatter settlements? (*6L, 5M*)

D Discuss the positive aspects of the quality of life in squatter settlements. (*6L, 5M*)

E With reference to a city you have studied, describe and explain the location of squatter housing. (*12L, 8M*)

107. Health costs of urban air particulate pollution

Figure 136 Projected health costs ($bn), 1995–2020

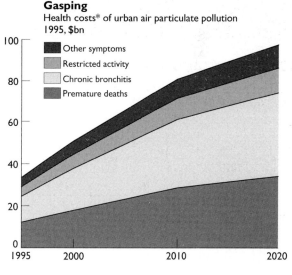

Source: World Bank.

*Social costs of mortality and morbidity due to exposure to PM10

A What are the main causes of particulate pollution in urban areas? (*5L, 3M*)

B What are the reasons for the considerable increase in projected health costs of urban particulate pollution? (*6L, 4M*)

C Suggest how the global distribution of such health costs is likely to change over the time period covered by the graph. (*6L, 4M*)

D Discuss the following quotation: 'indoor air pollution is inextricably linked with poverty, and should therefore become less of a problem as living standards rise'. (*The Economist*, 21/3/1998.) (*8L, 6M*)

E With reference to an urban area you have studied, discuss what has been done to tackle the air pollution problem. (*10L, 8M*)

108. Favelas and low-cost government housing in Brazil

Figure 137 Favela: edge of central São Paulo

A Describe the favela shown in Figure 137 which is on the edge of central São Paulo. (*4L, 3M*)

B Why are some favelas, like the one shown in Figure 137, found near the centre of cities in LEDCs? (*6L, 4M*)

C Suggest why such favelas are likely to be temporary in nature. (*6L, 4M*)

D Describe the low-cost government housing shown in Figure 138. (*4L, 3M*)

E Why is such housing so limited in number in Brazil and other LEDCs? (*5L, 3M*)

F With reference to an urban area in an LEDC you have studied, discuss what has been done to improve the housing situation. (*10L, 8M*)

Figure 138 Low-cost government housing in Manaus

109. New Towns

Figure 139 Crawley, West Sussex

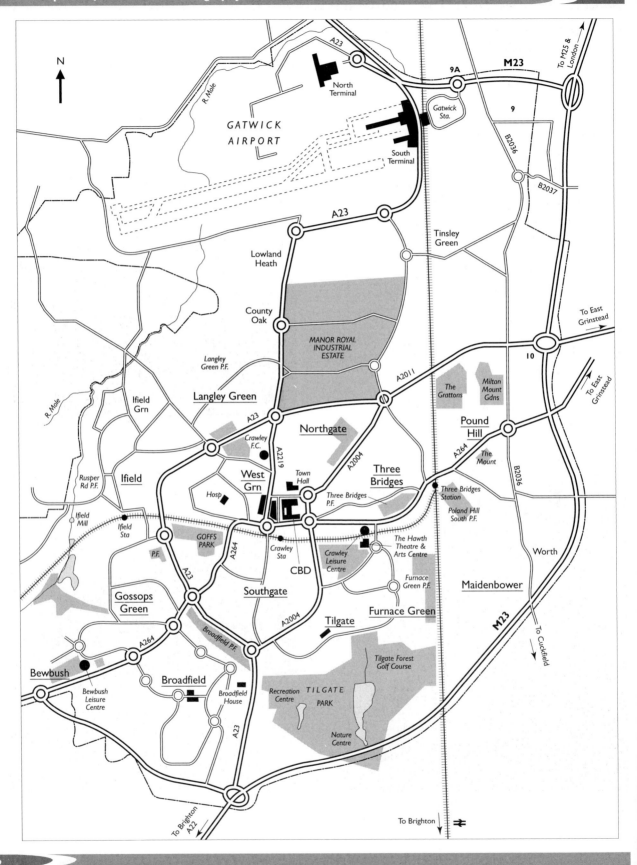

A Suggest and justify a location within Crawley for each of the five photographs. (*5L, 5M*)

B Describe two characteristics of New Towns. (*6L, 4M*)

C How might you expect the population composition of a New Town to differ from that of a long-established urban area? Give reasons for your answer. (*8L, 4M*)

D Describe and explain the location of New Towns in one region of the UK. (*6L, 4M*)

E Discuss the development of a New Town you have studied. (*10L, 8M*)

110. New patterns of retailing

A Suggest and justify two locations where the type of retail environment shown in Figure 140 is frequently found. (*6L, 4M*)

B For one of the locations you have identified in A, explain why such retail environments have grown in number in recent years. (*6L, 4M*)

C What is the usual name applied to the type of retail environment shown in Figure 141? (*1L, 2M*)

D Describe the characteristics of this type of retail environment. (*6L, 4M*)

E Where are such retail environments usually located? Explain why. (*6L, 4M*)

F Examine the characteristics of one other type of retailing that has expanded significantly in recent decades. (*10L, 7M*)

Figure 140 Example A: modern retailing

Figure 141 Example B: modern retailing

111. The Clark-Fisher sector model

Figure 142

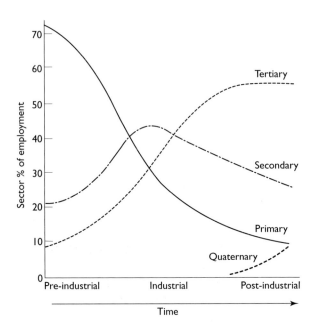

Figure 143 Maine, USA

Figure 144 Maine, USA

A Describe the sectors of the economy illustrated by Figures 143 and 144. (*6L, 4M*)

B Briefly describe the changes that occur in the three stages of the model (Figure 142). (*6L, 4M*)

C Why does the fall in primary employment slow considerably in the post-industrial stage? (*5L, 4M*)

D Account for the changes that occur in secondary employment from one stage to another. (*8L, 6M*)

E Explain the contrasting trends in tertiary and quaternary employment in the post industrial stage. (*10L, 7M*)

112. Transnational companies

Figure 145 The development of TNCs

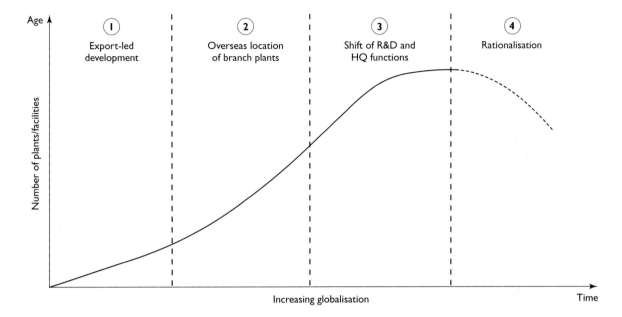

A Define the terms 'transnational company' and 'globalisation'. (2 × [*3L, 2M*])

B Suggest three reasons why some companies, when they reach a certain size, take the decision to become transnational. (3 × [*3L, 2M*])

C Describe and explain the four stages that the development of a typical TNC follows over time. (*6L, 4M*)

D Why are most governments generally keen to host foreign TNCs? (*6L, 4M*)

E What are the possible disadvantages of too high a level of TNC investment in a country? (*8L, 7M*)

113. The Rostow model

Figure 146 is an incomplete diagram of the Rostow model.

A What are the usual titles given to stages 3 and 4 of the model? (*2L, 2M*)

B Name an example of a country which approximates the conditions in

 (i) Stage 2 and

(ii) Stage 5. (*2L, 2M*)

C How would you expect the structure of manufacturing industry to differ between stages 2 and 5? (*6L, 5M*)

D Why has it proved very difficult for many developing countries to reach the later stages of the model? (*10L, 7M*)

E For one developed country you have studied, briefly discuss its development from Stage 1 to Stage 5. (*15L, 9M*)

Figure 146

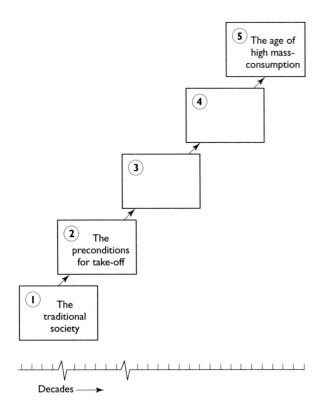

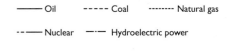

Figure 147

(a) Production of energy in the United Kingdom

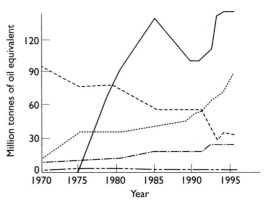

(b) Decline in coal production in the United Kingdom

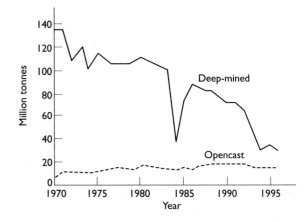

114. The decline of Britain's coal industry

Figure 147 shows energy production and coal-mining in the UK.

A Describe the changes in energy production in the UK between 1970 and 1995. (*5L, 3M*)

B Discuss three factors that have resulted in the sharp decline in coal production. (*8L, 6M*)

C Why has the decline in employment in the coal industry been much greater than the drop in coal production? (*5L, 4M*)

D Examine the impact of such changes on mining communities. (*7L, 5M*)

E What can the government do to help distressed mining communities? (*10L, 7M*)

(c) Decline in coal mining employment in the United Kingdom

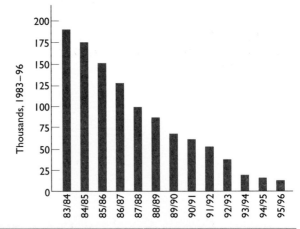

115. OECD employment trends

A Explain the technique used in the diagrams to show changes in employment. (*4L, 3M*)

B Describe the changes in manufacturing employment in the OECD between 1985 and 1995. (*5L, 4M*)

C Suggest one factor which could

(i) account for the increase in manufacturing employment between 1985 and 1990

(ii) explain the fall in manufacturing employment after 1990. (2 × [*4L, 2M*])

D To what extent did total employment in services change during this period? (*6L, 6M*)

E Why did some service sectors increase in employment much faster than others? (*12L, 8M*)

Figure 148 OECD employment trends in manufacturing and service industries

116. Cambridge: a high-technology cluster

A What is a 'high-technology cluster' (Figure 149)? (*4L, 3M*)

B Suggest reasons for the clustering of high-technology in and around Cambridge. (*8L, 5M*)

C Why is new investment required for the high-technology cluster to reach its full potential? (*10L, 7M*)

D Assess the costs and benefits of significant further development in this region. (*13L, 10M*)

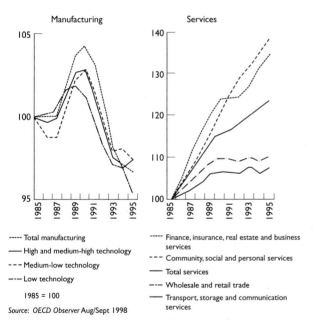

Manufacturing | Services

----- Total manufacturing

—— High and medium-high technology

--- Medium-low technology

—·— Low technology

1985 = 100

Source: OECD Observer Aug/Sept 1998

----- Finance, insurance, real estate and business services

--- Community, social and personal services

—— Total services

—·— Wholesale and retail trade

—— Transport, storage and communication services

Figure 149 The Cambridge high technology cluster

CAMBRIDGE:
Signpost to the future?

NEXT WEEK, a consultants' report will land on the desks of some of East Anglia's most influential business executives and politicians, including Sir Alec Broers, Vice-Chancellor of Cambridge University. Its contents are so important to the future of the British economy they will also be sent to 10 Downing Street. Commissioned by Government, the report will add weight to the position of the Greater Cambridge Partnership, a high-level promotional group which believes the authorities should be "investing in success".

Instead of just helping the regeneration of poor communities, the argument runs, the Government – with private-sector support – should pour billions into some of the fastest-growing and most prosperous regions in the country. This has triggered some soul-searching within New Labour, whose instincts are to leave economic development to the markets and focus Government resources on education, health and tackling poverty. More specifically, the report will back the view that the Greater Cambridge hi-tech cluster – Silicon Fen – should be at the top of the Government's list when it comes to handing out development funds to already prosperous areas. If this region's potential is to be unleashed, it will say, urgent action is required.

More than £2 billion of new money needs to be mobilised over the next five to 10 years, some of it through public private partnerships, to provide social and economic infrastructure – better roads, railways, schools and thousands of new houses.

Without such investment, the city – home to semiconductor designer ARM Holdings, software engineer Autonomy and biotech leader Cambridge Antibody Technology – will be suffocated by shortages of skilled and semi-skilled labour, an inadequate transport system and lack of housing. Worse, says Hermann Hauser, the area's leading venture capitalist, this will happen just as Cambridge has achieved the critical mass which could ensure it a vital role in the future growth of Britain's knowledge economy.

But in order for investment on this scale to be put into place swiftly, a series of highly controversial political decisions will have to be taken. The laborious planning procedures and often-conflicting structures of local government will have to be subordinated to an economic strategy agency with explicit planning authority, local officials say.

In a hint of the upheavals that lie ahead for its unsuspecting, predominantly rural, communities, they recognise the time is fast approaching when it will be necessary to encroach on the strict green belt which protects the beautiful medieval university town. The population of Cambridge has been effectively limited by planning directives, dating from the late 1940s, to around 110,000. Now a new settlement needs to be built, ultimately to provide housing for up to 25,000 located on the edge of the green belt.

Because the rapid growth of Cambridge's hi-tech cluster has sent housing prices soaring towards London levels, even eminent academics are now turning down offers of professorships. Provision must therefore be made for homes which university lecturers, teachers, nurses, laboratory technicians and software engineers can afford.

If the correct decisions are taken now, its supporters say, Cambridge's tech cluster could become as important a source of wealth and innovation for Britain as California's Silicon Valley has been to America or the financial sector of the City of London is now to Britain.

Source: Evening Standard 6/9/01

117. Foreign direct investment

Figure 150 shows global shares of foreign direct investment.

Figure 150

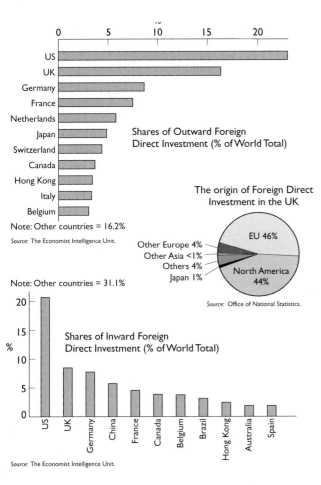

Shares of Outward Foreign Direct Investment (% of World Total)

Note: Other countries = 16.2%

Source: The Economist Intelligence Unit.

The origin of Foreign Direct Investment in the UK

Note: Other countries = 31.1%

Shares of Inward Foreign Direct Investment (% of World Total)

Source: The Economist Intelligence Unit.

A Describe and explain the origin of global outward foreign direct investment. (*5L, 3M*)

B Why do countries such as the USA and the UK invest so much abroad? (*5L, 3M*)

C To what extent and why does the global share of inward foreign direct investment differ from that of outward foreign direct investment? (*7L, 5M*)

D Suggest reasons for the origin of foreign direct investment in the UK. (*8L, 6M*)

E What are the advantages and disadvantages for a country of a high level of inward foreign direct investment? (*10L, 8M*)

118. Global cities

Figure 151 shows the network of global cities.

Figure 151

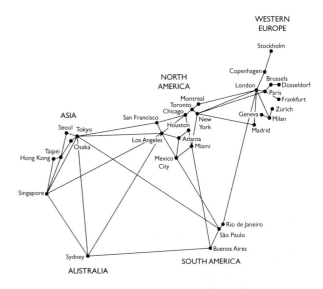

A What do you understand by the term 'global city'? (*4L, 3M*)

B Describe the distribution of global cities. (*5L, 4M*)

C Discuss the reasons for this distribution. (*6L, 4M*)

D Suggest how this distribution might change over the next 20 years or so. (*6L, 6M*)

E For one of the cities shown in Figure 151, explain the reasons for its designation as a global city. (*14L, 8M*)

119. The impact of plant closure

Figure 152 The decline of the UK manufacturing industry 1978–2000

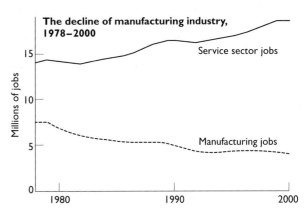

Figure 153 The closure of a large manufacturing plant in the UK

Vauxhall closure

CAR workers in Luton can expect advice and sympathy but no Government rescue plan to save the Vauxhall plant.

Although unions described yesterday's announcement by General Motors as a "bolt out of the blue" it evidently did not come as a surprise to ministers, who played down all attempts to draw a political message from what they say is an unavoidable restructuring of the car industry.

Within five minutes of the announcement that car production in Luton was to cease, Stephen Byers, the Trade and Industry Secretary, announced a range of measures to soften the impact on the local economy and help former car workers into other employment.

The Tories blamed the closure on Labour's tax policies, while the pro-euro lobby, including several union leaders, attributed the closure to the high pound compared with the euro. Government spokesmen privately dismissed both theories.

Tony Blair described the announcement as "very bad news" but pointed to the fact that 2,000 out of the total of 5,000 jobs lost would be in Germany as evidence that this was not a peculiarly British problem.

The Government's moves yesterday contrast starkly with its reaction when BMW threatened to pull out of Longbridge last year. Then Mr Byers did not hide his anger with the German company, and made strenuous efforts to keep production going at the plant.

He said yesterday: "Vauxhall's announcement is a bitter blow for the individuals affected, their families and the community. Our key aim will be to find new job opportunities to replace those being lost over the next year, and the Government is determined to play its part."

He added: "Despite the disappointing news, Vauxhall remains an important manufacturer in the UK, employing over 8,000 people. Vauxhall has confirmed that it will continue with its investment plans for a new van to be manufactured at Luton."

He promised help via the Regional Development Agency for companies which supply the Luton plant, and a package of advice and assistance for workers who are made redundant.

The Employment Service is to send in its rapid response units to give personal service to those looking for jobs. The Government will also put up loans to fund new business ventures.

Source: Daily Telegraph 13/12/2000

A Describe the changes in manufacturing and service sector employment in the UK between 1978 and 2000. (Figure 152) (*4L, 3M*)

B (i) Give two service industries where employment has increased in recent decades.

(ii) Give two manufacturing industries which have lost jobs in the last 20 years. (*2 × [2L, 2M]*)

C Discuss the reasons for the closure of Vauxhall's Luton car plant. (*6L, 4M*)

D Examine the potential consequences of the closure of this plant. (*9L, 6M*)

E What can the government do to minimise the impact of the closure of large plants such as Vauxhall in Luton? (*12L, 8M*)

120. The geography of poverty and wealth

Figure 154

The Geography of Poverty and Wealth

Interpreting the Patterns

In our research we have examined three major ways in which geography affects economic development. First, as Adam Smith noted, economies differ in their ease of transporting goods, people and ideas. Because sea trade is less costly than land- or air-based trade, economies near coastlines have a great advantage over hinterland economies. The per-kilometer costs of overland trade within Africa, for example, are often an order of magnitude greater than the costs of sea trade to an African port.

Second, geography affects the prevalence of disease. Many kinds of infectious diseases are endemic to the tropical and subtropical zones.

According to the World Health Organization, 300 million to 500 million new cases of malaria occur every year, almost entirely concentrated in the tropics. The disease is so common in these areas that no one really knows how many people it kills annually – at least one million and perhaps as many as 2.3 million. Widespread illness and early deaths obviously hold back a nation's economic performance by significantly reducing worker productivity. But there are also long-term effects that may be amplified over time through various social feedbacks.

For example, a high incidence of disease can alter the age structure of a country's population. Societies with high levels of child mortality tend to have high levels of fertility: mothers bear many children to guarantee that at least some will survive to adulthood. Young children will therefore constitute a large proportion of that country's population. With so many children, poor families cannot invest much in each child's education. High fertility also constrains the role of women in society, because child rearing takes up so much of their adult lives.

Third, geography affects agricultural productivity. Of the major food grains – wheat, maize and rice – wheat grows only in temperate climates, and maize and rice crops are generally more productive in temperate and subtropical climates that in tropical zones. On average, a hectare of land in the tropics yields 2.3 metric tons of maize, whereas a hectare in the temperate zone yields 6.4 tons. Farming in tropical rain-forest environments is hampered by the fragility of the soil: high temperatures mineralize the organic materials, and the intense rainfall leaches them out of the soil. In tropical environments that have wet and dry seasons – such as the African savanna – farmers must contend with the rapid loss of soil moisture resulting from high temperatures, the great variability of precipitation, and the ever present risk of drought. Moreover, tropical environments are plagued with diverse infestations of pests and parasites that can devastate both crops and livestock.

Moderate advantages or disadvantages in geography can lead to big differences in long-term economic performance. For example, favorable agricultural or health conditions may boost per capita income in temperate-zone nations and hence increase the size of their economies. This growth, encourages inventors in those nations to create products and services to sell into the larger and richer markets. The resulting inventions further raise economic output, spurring yet more inventive activity. The moderate geographical advantage is thus amplified through innovation.

In contrast, the low food output per farm worker in tropical regions tends to diminish the size of cities, which depend on the agricultural hinterland for their sustenance. With a smaller proportion of the population in urban areas, the rate of technological advance is usually slower. The tropical regions therefore remain more rural than the temperate regions, with most of their economic activity concentrated in low-technology agriculture rather than in high-technology manufacturing and services.

We must stress, however, that geographical factors are only part of the story. Social and economic institutions are critical to long-term economic performance.

The Wealth of Regions

Climate Zone (percent of world total)		Near*	Far*
Tropical			
Land area	19.9%	5.5%	14.4%
Population	40.3%	21.8%	18.5%
GNP	17.4%	10.5%	6.9%
Desert			
Land area	29.6%	3.0%	26.6%
Population	18.0%	4.4%	13.6%
GNP	10.1%	3.2%	6.8%
Highland			
Land area	7.3%	0.4%	6.9%
Population	6.8%	0.9%	5.9%
GNP	5.3%	0.9%	4.4%
Temperate			
Land area	39.2%	8.4%	30.9%
Population	34.9%	22.8%	12.1%
GNP	67.2%	52.9%	14.3%

* "Near" means within 100 kilometres of seacoast or sea-navigable waterway; "far" means otherwise.

Source: Scientific American, March 2001

A Describe the relationship between population and GNP for the four regions shown in Figure 154. (*5L, 4M*)

B To what extent does this relationship differ between the 'near' and 'far' areas of each region? (*6L, 4M*)

C Discuss the reasons for the differences you have described in B. (*6L, 4M*)

D Explain the geographical advantages that the temperate climatic zone has over the other three regions. (*8L, 5M*)

E Why are 'social and economic institutions critical to long-term economic performance'? (*10L, 8M*)